Volker Simmering

The Evolution of Standards

GABLER EDITION WISSENSCHAFT

Ökonomische Analyse des Rechts

Herausgegeben von
Professor Dr. Peter Behrens
Professor Dr. Manfred Holler
Professor Dr. Claus Ott
Professor Dr. Hans-Bernd Schäfer (schriftführend)
Professor Dr. Rainer Walz
Universität Hamburg, Fachbereich Rechtswissenschaft II

Die ökonomische Analyse des Rechts untersucht Rechtsnormen auf ihre gesellschaftlichen Folgewirkungen und bedient sich dabei des methodischen Instrumentariums der Wirtschaftswissenschaften, insbesondere der Mikroökonomie, der Neuen Institutionen- und Konstitutionenökonomie. Sie ist ein interdisziplinäres Forschungsgebiet, in dem sowohl Rechtswissenschaftler als auch Wirtschaftswissenschaftler tätig sind und das zu wesentlichen neuen Erkenntnissen über Funktion und Wirkungen von Rechtsnormen geführt hat.

Die Schriftenreihe enthält Monographien zu verschiedenen Rechtsgebieten und Rechtsentwicklungen. Sie behandelt Fragestellungen aus den Bereichen Wirtschaftsrecht, Vertragsrecht, Haftungsrecht, Sachenrecht und verwaltungsrechtliche Regulierung.

Volker Simmering

The Evolution of Standards

With a Foreword by Prof. Dr. Manfred J. Holler

Deutscher Universitäts-Verlag

Bibliografische Information Der Deutschen Bibliothek
Die Deutsche Bibliothek verzeichnet diese Publikation in der Deutschen
Nationalbibliografie; detaillierte bibliografische Daten sind im Internet über
<http://dnb.ddb.de> abrufbar.

Dissertation Universität Hamburg, 2002

1. Auflage Mai 2003

Alle Rechte vorbehalten
© Deutscher Universitäts-Verlag GmbH, Wiesbaden, 2003

Lektorat: Brigitte Siegel / Sabine Schöller

Der Deutsche Universitäts-Verlag ist ein Unternehmen der
Fachverlagsgruppe BertelsmannSpringer.
www.duv.de

Umschlaggestaltung: Regine Zimmer, Dipl.-Designerin, Frankfurt/Main

Gedruckt auf säurefreiem und chlorfrei gebleichtem Papier

ISBN-13:978-3-8244-7832-3   e-ISBN-13:978-3-322-81514-9
DOI: 10.1007/978-3-322-81514-9

Foreword

The first essay in Volker Simmering's study in *The Evolution of Standards*, undertaken while at the *Graduate Programme in Law and Economics* in cooperation with the *Institute of SocioEconomics* at the University of Hamburg, examines the effects of "globalization" on the evolution of global standards. The central result of this essay is that there may be too few global standards and those which do evolve are not necessarily efficient; a result that is at odds with standard economic reasoning. Simmering derives his result from the application evolutionary theory – which is becoming increasingly popular in economics, the social sciences, and philosophy. This branch of theory focuses on the convergence and selection of equilibria and its attendant reduction of variety.

Simmering's study, very interestingly, indicates a major difference in the results of evolutionary reasoning when applied to the coordination of human interaction to that at the species level. Simmering shows how variety is reduced to produce homogeneity or compatibility; while in biology it has been to show how variety emerges. For example, in his *The Origin of Species* (1859), Charles Darwin tried to give an interpretation of the characteristics, diversity, and distribution of the various forms of animal and plant life as the result of a historical process involving descent with modification. This process includes the origin of two or more distinct species, or other groups, from a single ancestral group, and also the natural modification of a line of ancestors and their descendents to such a degree that the later forms may be markedly different from the earlier. The question in economics is different. It is not to explain why so much variety exists but whether an equilibrium that is selected by a process of evolution is efficient; or whether the intervention of benevolent dictator can improve the outcome from a social point of view.

The message of Herbert Spencer's Social Darwinism and its trivialization as "survival of the fittest" suggests that society progresses to efficiency. However, fitness is defined relative to the surrounding conditions, *i.e.* the property of relevant populations. But, even nature is not always "successful" in selecting a single best alternative. For instance, a large variety of species of antelopes graze side by side in the Serengeti. Their horns are evidently useful and yet may be of different forms in different species. It seems impossible that the characteristic form in each species is precisely that best fitted for the survival of that particular species.

Among other concepts of evolutionary game theory, Volker Simmering builds upon an approach of Ralf Peters (1998)[1], also developed at the Institute of SocioEconomics, and applies the size of the basin of attraction to discuss stability and convergence properties of his industry model. Through this, he derives interesting conclusions for the optimal decision on the legal status of formal standards.

Simmering's second essay analyzes conditions of technological progress in networks under the assumption of switching costs and a limited life span for goods. Users have to decide whether they exhaust this life span or prematurely switch to a technologically advanced version of the good. The rational solution to this problem depends clearly on what others do. Simmering demonstrates that, counter-intuitively, the presence of switching costs tends to support transition to new technologies under these conditions. The third essay derives from joint work on harmonization within the European Union by means of majority voting in

---

[1] Peters, R. (1998): Evolutionäre Stabilität in sozialen Modellen. (Diss.) München.

committees such as CEN and CENELEC (Holler and Simmering, 2001)[2]. Voting outcomes depend on whether the components of a system good are standardized so that they are compatible and, *e.g.*, the amplifier of system A can be used in combination with the loudspeakers of system B. In case of standardization, harmonization on a mix of components of systems A and B is possible and a majority voting on various combinations to single out a unique mix of components is appropriate. However, anticipation of the voting outcome, Simmering shows, may cause producers to keep their systems incompatible. This implies that the optimal system may not be available for harmonization through voting.

Simmering's book will be of interest to scholars of standardization and network economics. The results of his last essay contribute to the understanding of European Standardization. The results of the second essay are potentially relevant for decision making on the supply and demand of technologically advanced goods and thus they contribute to the explanation of technical progress. The evolutionary model in the first essay should motivate a more profound understanding of socio-economic processes and solutions to coordinating social interaction. Simmering's message is key to critically assess the more standard perspective of rational choice.

Manfred J. Holler

---

[2] Holler M.J. and Simmering, V. (2001): "Voting on Harmonization." Paper presented at the Annual Meeting of the European Public Choice Society, Paris, 2001.

Preface

I wish to use this opportunity to thank the people and the institutions that made this dissertation possible. I begin with thanking my doctoral supervisor, Prof. Manfred Holler, for his valuable advice and support that greatly benefited both me, personally, and my thesis. My sincere thanks goes to Prof. Hans-Bernd Schäfer for his role as the director of the Graduate College for Law and Economics and the second supervisor of my thesis. In his role as the former, his great dedication to setting up and shaping the Hamburg Doctoral Program in Law and Economics, making it a unique institution and an outstanding chance for its participants, is commendable. As my co-supervisor, he has been instrumental in introducing me to new ideas. I next thank Prof. Norbert Schulz for supporting and encouraging me. I also thank Prof. Robert Cooter for making my stay at the UC Berkeley as a visiting scholar possible, which was a great experience and help for my research.

I thank Markus Berndt and Padmashree Gehl Sampath as well as Heide Coenen, Prof. Alfred Endres, Helge Janßen, Prof. Marc Lemley, Ines Lindner and Dirk Reuscher for their helpful comments and inspiring discussions. I thank Oliver Goettsche for his patience and his careful work on the manuscript.

I gratefully acknowledge financial support by the German National Research Foundation (DFG) and the Berkeley Law and Economics Program.

Finally, I thank my parents and Ulrike Hahne for their love, encouragement and help.

Volker Simmering

Abstract

The thesis comprises three essays on the evolution of standards in networks. Networks exhibit positive adoption externalities, so-called "network effects". The essays investigate various aspects of how such effects influence the performance of markets and other institutional arrangements.

In the first essay, an evolutionary game theoretic model is introduced in order to study how increasing integration among nations ("globalization") influences the evolution of global standards. It is found that integrated markets tend to produce too few global standards. Even if global standards do evolve, agents do not necessarily end up with the best available one. In addition, strategies for policy intervention are discussed. The role of laws that render adoption of technical standards mandatory are compared with voluntary standards issued by recognized standardization bodies.

The second essay is devoted to the problem of technological progress in networks. It focuses on industries where the goods that are associated with each technology have a limited life and users, due to sunk costs involved, are either more or less committed to their chosen technology. The study suggests that the risk of too much technological change is low in such industries. In contrast, users tend to suffer from too little technological change. The lower the users' commitments the higher the risk that established technologies are too sticky. A fruitful strategy of governmental technology policy, the analysis suggests, is to fasten the transition process if an already ongoing transition is observed.

The third essay studies the resolution of conflicts within international standardization bodies. We discuss whether voting is an eligible mechanism for collective decision making in such bodies. It is demonstrated that outcomes of votes on harmonization of international standards are likely to be stable and not path dependent, even if goods are specified in more than one dimension. In addition, pitfalls that might arise with harmonization policy are identified. In case the harmonization body cannot commit to abstain from harmonization – even if it has observed that the industry performs badly – the outcome with harmonization policy might be inferior to the outcome that would be achieved without harmonization policy. Thus, *ex post* desirable harmonization policy might be undesirable *ex ante*.

.

Table of Contents

1. Introduction

The first human beings that suffered from a lack of standards might have been the residents of Babel. In order to stop their ambitious attempt to build a tower that reaches the heavens, God made them speak different languages. Since verbal communication proved to be an essential coordination facility in Babel, their project failed disastrously and the people from Babel dispersed throughout the world. This story illustrates an important property that languages possess, which is also shared by many other goods: their "value" is positively correlated to the number of people who adopt the same "goods" – in other words, they exhibit "network effects". Such adoption externalities play a major role in modern life. The value of an Internet portal, a specific type of music player as well as many social and legal (coordination) norms and other institutions, for example, heavily depends on the (expected) number of people that adopt the *same* or *compatible* "goods". Since thus, in turn, individuals' benefits may crucially depend on other players making the same (or a compatible) choice, the evolution of standards is of vital importance in these areas.

The story of the Tower of Babel also suggests that there might be processes that facilitate the evolution of standards. After all, before God intervened, the people had shared a common language. Nonetheless, people today often complain about a lack of standards in various areas. For example, a recent article in the *Spiegel*, headlined "Virulent Diversity", presents a number of examples where consumers suffer from of a "lack" of standards for plugs and sockets, printer cartridges and drivers, mobile phone chargers, coffee filter, vacuum cleaner bags, SCSI- and SCART-interfaces, etc.[1] It is, however, clear that people often do benefit from variety, even if network effects are involved. Should the advertising community give up its beloved Macintosh computers? Do we really want a universal world language? Thus, the presence of network effects generates a tradeoff. While harmonization enhances network benefits, it also diminishes the benefits derived from variety. However, due to the externalities involved, one may become very skeptical as to whether markets actually implement the optimal solution to this tradeoff; and if not, do markets produce too much or too little harmonization?

Technological change is yet another problem consumers potentially face in networks. In order to be superior, quite often, technologies have to be designed incompatible to established ones. Consider, for example, the CD-player as a technological advancement to the classical record player. On the one hand, one might assume that users hesitate excessively when faced with the option of switching to new technologies because those who switch early do not want to relinquish network benefits. Conversely, however, network effects are particularly relevant for durable goods. Thus, if transition occurs, adopters of the new technology impose potentially severe losses on "stranded" users. How do these effects trade off? Do we have too much or too little technological revolutions in networks?

Suppose that network effects do cause the market to systematically fail. Should policy- and/or lawmakers intervene? If so, which policy strategies are likely to be successful? Do we need mandatory rules and/or subsidies in such areas? Are there alternative institutions that better support the optimal evolution of standards?

These and related questions have been studied by a growing number of researchers. Following early works by David (1985), Farrell and Saloner (1985) and Katz and Shapiro (1985), the economics of networks and standardization has become a widely recognized research area. Comprising three essays on various aspects on the evolution of standards, this thesis aims at deepening further our understanding of how network effects influence the functioning of markets and other institutional arrangements. Each of the three essays studies both "pure" markets and options of policy- and lawmakers that are deemed to support the optimal evolution of standards.

As of yet, little attention has been devoted to the explicit analysis of the evolution of standards within an international context, which is the subject of the first essay. Due to the sharply reduced costs of moving goods, money, people, and information, as well as through relaxations of governmentally established "separation fences" such as tariffs and export subsidies, many countries have become more and more integrated.[2] It is obvious that this development diminishes the network benefits that purely national standards provide. Since, as argued above, the optimal degree of harmonization is a solution to the tradeoff between network benefits and benefits from variety, increasing integration ("globalization") intensifies the need for common international standards. This leads us to the following questions: Do markets react efficiently to globalization? Does globalization produce too much or too little harmonization of national standards? Do increasingly integrating international markets yield harmonization too early or too late? If globalization does lead to harmonization, does it imply that the "right" standard is implemented?

In order to answer these questions we introduce an evolutionary flavored model, which is based on pairwise interactions of agents originating from two large populations, interpretable as countries. In each interaction, agents face a coordination game where both players' payoffs are always higher if both choose the "same" strategy. The choice of strategy can be inter-

---

[1]   See the SPIEGEL of 23rd March 2001 (translations from German by the author).
[2]   See Sykes (1995), p. xii.

preted as the adoption of a particular network good such as a specific social norm, a particular language, some type of software, etc. An exogenous integration parameter is included, in order to analyze the impact of globalization.

Our analysis suggests that globalization does not necessarily implement harmonization of national standards even if it would be efficient. Even *if* globalization does implement harmonization it does so too late from a social point of view. Thus, people tend to suffer from a lack of global standards. Moreover, we find that globalization may produce harmonization with an inferior global standard, *i.e.* the domestic standard of country *i* may very well prevail as the international one, even though harmonization with the domestic standard of country *j≠i* would produce higher utilitarian welfare. In an extension of the model, it is also shown that the possibility of adopting more than one standard (such as learning several languages) supports efficient harmonization, even if such double adoptions do not occur in stable equilibrium.

Since the analysis suggests that populations may get stuck in an inefficient "variety" state, the second part of this essay deals with options available to policy-makers to implement harmonization. Two basic strings of eligible measures are analyzed: a) mandatory standards that are supplemented by sanctions and b) voluntary standards that are issued by recognized standardization bodies, which are non-enforced, yet may still trigger an initial collective switch. As long as voluntary standards suffice to implement harmonization (*i.e.* the "stability" of variety is low), it is likely that they are preferable to mandatory standards because they avoid costly enforcement activities and administration costs among other costs.

We find that harmonization policy in the EU does, in fact, make use of both kinds of measures. For many harmonization projects it relies on standards issued by non-governmental standardization bodies. Through granting an "official" and "exclusive" status to selected standardization bodies, EU policy-makers strengthen the "collective switch power" of the standards issued. In addition, non-adoption of some of these standards is either directly or indirectly sanctioned. For example, the adoption of some standards is declared mandatory for specific applications. Besides, the adoption of other standards that are linked to a specific harmonization directive and are mandated by the European Commission insure compliance to national regulations. This is likely to serve as an "indirect" sanction for the non-adoption of these standards. It is further argued that such a procedure entails the problem that budget maximizing "bureaucrats" within the official standardization bodies may have excessive incentives for harmonization. To reduce these bodies' "harmonization power" is, however, not an immediate solution to this problem, since then bureaucrats might too often choose an inferior candidate for harmonization.

The second essay deals with technological progress in networks. Often, the users who first switch to a new network technology are initially worse off, even though, given the same number of adopters, the new technology is superior. For example, the first people who replaced their analogue record player with a CD-player suffered from a lack of variety in music discs. The essay analyzes whether and under which conditions such new (statically) superior network technologies prevail over inferior established ones. If we assume that the goods pertaining to either technology are competitively supplied, then are established network technologies too sticky ("excess inertia") or are they replaced too often ("excess momentum")?

Crucial for a new network technology's success over an established one is that users are sufficiently committed to their adopted technology. If commitment is too low, users lack incentives to incur initial losses associated with an early switch to the new technology. Users would rationally pursue a "jump-on-the-bandwagon" strategy. Since every agent shares this incentive, the bandwagon would never start rolling in the first place. Only if agents are sufficiently committed to the chosen technology, could they have incentives to switch to an emerging technology, even if their payoffs are lower at the beginning. Their commitments are responsible for the agents' fear to of abandonment, which may make them willing to incur these initial losses. This might get the bandwagon rolling. On the other hand, however, the stronger the users' commitment, the longer and the more costly the transition process becomes, which diminishes the incentives to switch to an emerging technology.

To verify these ideas, we set up a dynamic game-theoretic model, which, contrary to the models in the established literature, allows for varying commitment levels, which correspond to the (sunk) costs to be incurred to acquire a technology's good (*i.e.* switching costs). We focus on industries where the life-span of goods pertaining to each technology is limited. Thus, agents are uncommitted from time to time. We find that transition to the new technology occurs if agents who are uncommitted after introduction of the new technology are willing to switch to the new technology. Whether or not these agents are willing to switch, depends on two conditions. First, their payoff stream must be larger when transition occurs than when the population stays with the established technology. Second, these agents' payoffs should be larger if they switch immediately rather than at any later date, *i.e.* these agents may not have an incentive to pursue a jump-on-the-bandwagon strategy.

Afterwards, it is asked how these conditions relate to the social desirability of transition. For the specific, often applied, case of linear network benefits functions (and equal distribution of reinvestment dates), we find that "maximum" commitment implies that excess momentum cannot occur for any set of parameter values. Even with maximum commitment, users may suffer from excess inertia; however, transition to the challenging technology does occur if every agent prefers transition. For commitment values that lie moderately below maximum commitment, transition is more expeditious, more likely and more desirable. However, as the desirability of transition increases at an even greater rate than "early" agents' incentives to switch, the risk of excess inertia increases. For even lower commitments, there exists some critical level below which the early agents' incentive to jump on the bandwagon becomes binding. Further decreases in commitments lessen the likelihood of transition, even though the transition time would be further reduced and the desirability of transition would be increased. No matter how superior the challenging technology is, a commitment level always exists below which transition undesirably fails.

The essay also deals with the transition *time*. Even though the transition process is shorter if commitment decreases, commitment levels below maximum imply that the transition time is always too long. One reason for this is that agents who are still committed to the old technology are rationally apathetic. That is, they are not able to coordinate a "premature" simultaneous switch to the new technology, even though such a collective action would make every user better off. If the transition time was shortened, not only would the social benefits of transition be enhanced but also the problem of excess inertia would be reduced.

Finally, the question is posed, whether policy-makers should intervene. Since the risk of excess momentum is limited in the model, policy-makers could first wait and see whether new technologies succeed without them interceding. It is further proposed that the risk of excess inertia should be reduced through an exogenous acceleration of the transition process. Such a measure is especially fruitful, if policy-makers only focus on new network technologies whose diffusion process has already started.

The third essay is devoted to the work of standardization bodies and the resolution of conflicts within them. In the previous essays, several potential market failures associated with network effects have been identified, and we have analyzed whether intervention is recommendable and, if so, how policy-makers or some other authorized body should go about it. In considering such a body as an entity that strives to achieve some well-defined goal, we have ignored the decision process *within* that body. In fact, in many international standardization bodies conflicts are likely to emerge. The reason is that the members of such bodies' usually correspond to the "official" national standardization bodies. One may suppose that these national bodies, at least to some degree, act in the interest of the country they represent. Even if every country prefers international harmonization of standards, it seems clear that different countries often favor different standards – often their "domestic" one – for common adoption. To resolve such conflicts, most of the major international standardization bodies[3] apply

---

[3]   For example, the CEN and CENELEC are the major European "official" standardization bodies; ISO and IEC are recognized international bodies.

voting. Is voting an appropriate mechanism for collective decision-making in such bodies?

The essay focuses on system goods. Since such goods are specified in more than one dimension, voting is inherently concerned with the risk of cyclical majorities. It is, however, argued that cyclical majorities do not necessarily evolve, because the "product space" is likely to reduce to a discrete grid, which implies that there may be a winning proposition that is chosen independently from the voting agenda. Thus, bargaining and conflicts about agenda setting are a secondary problem, which is favorable to the functioning of formal international harmonization through voting.

Another focus is strategic behavior under harmonization policy. Firms, in deciding whether to make systems' components combinable (vertically compatible), can manipulate the product space on which the standardization body's vote is based. Given that such a body cannot commit itself to abstain from harmonization – even if it observes that the market performs badly – harmonization policy might have a deleterious effect on the outcome rather than correcting the market failure. Thus, *ex post* optimal harmonization policy might be undesirable *ex ante*.

The thesis is structured as follows. In section 2, a brief introduction to the concept of network effects is given and we present established results and insights that are relevant for the rest of the study. Section 3 and 4 comprise the first essay on international harmonization. In section 5, the reader finds the essay on technological advance in networks. Section 6 contains the third essay, the analysis of voting on harmonization. Finally, section 7 briefly summarizes the main results of this thesis.

2.   The Economics of Networks, Compatibility and Standardization: Definitions, Basic Concepts and Insights

This chapter gives a brief introduction into the economics of networks and explains some concepts, insights and definitions, which we refer to in subsequent sections. First, the central concept of network effects is outlined in section 2.1 and it is briefly discussed how these effects potentially affect the performance of competitive markets (section 2.2). In section 2.3, we define the notions of compatibility, standardization and harmonization as they are used in this study. Section 2.4 briefly characterizes the role of switching costs, as far as they are relevant for the study, and attempts to outline their role in network markets. Section 2.5 deals with producers' incentives for establishing compatibility. Finally, section 2.6 briefly summarizes the results of the relevant empirical studies.

2.1   The Concept of Network Effects

Sometimes, the benefits an agent derives from the consumption of a durable good are positively correlated with the overall number of agents consuming the same or compatible goods. In turn, consuming such a good not only confers benefits on the buyer himself but also on other buyers. These positive effects a buyer bestows on other buyers are commonly referred to as network effects.[4]

Underlying nature

The underlying nature of (direct) network effects can be demonstrated using the metaphor of actual networks. Classic examples of actual networks are communication networks like e-mail, telephone or facsimile. The greater the number of people who possess an e-mail account (telephone access, fax-machine) the greater is the value they assign to being part of this network. Each new participant adds a new node and thus new links to an existing network, thereby the number of "goods" (links to communicate with other participants) the network provides to *all* participants is increased.[5] Communication, in general, is a composite good consisting of two strong complementary components, which are the network participation of *both* agents who (wish to) communicate with each other.[6] If there is a positive probability for each agent to desire communication via the network with each participant (*i.e.* participants ex ante benefit from variety), each additional participant yields additional benefits for *all* partici-

---

[4]   The concept of network effects entered economics research at the beginning of the 1980s. Pioneering works are Farrell and Saloner (1985, 1986a, 1986b) and Katz and Shapiro (1985, 1986).

[5]   This property of communication networks was pointed out by Rohlfs (1974).

[6]   See also Economides (1996).

pants. Thus, the value an individual assigns to have *access* to the network (through e-mail account, telephone access, fax machine) increases if the demand for *network accesses* increases.[7]

Three sources are in charge of direct network effects in communication networks. First, communication is a composite good. To be successful, communication requires that *both* parts, the sender and the receiver, join the same (or a compatible) network. Second, network accesses are durable goods. And third, at the time of adopting such a durable good, one does not yet know with whom one will make communication. Consequently, when joining a communication network, the *expected* number (or share) of people that will join the same (or a compatible) network is a crucial decision parameter.

A slightly different concept is that of *indirect network effects*, which may occur within markets for systems.[8] Systems are goods that consist of perfectly complementary components such as hardware and software[9]. Assume consumers benefit from the variety of systems. If economies of scale are involved in production of one or both components, each extra consumer may potentially increase the *number of variants* available. Thus, each buyer of a system confers a positive effect on other buyers.[10] A good example is computer hardware and software. In general, computer users strongly benefit from variety of software – hardware alone, without any software to run it, is often useless. "In fact, one can say that in such markets, once a consumer is committed to purchasing a specific brand, his / her satisfaction is derived only from the variety of the (specific) brand supporting service" (Chou and Shy 1996, p. 310). Firms usually face economies of scale with production of software – marginal costs typically come close to zero. Often, the production of a particular software program is only profitable if demand passes some threshold. Consequently, the *variety* of software may increase as the demand for systems increases.[11] Note that these effects occur even if marginal costs and prices for software are constant. As in the case of (direct) network effects, additional network participants (buyers of computer systems, owners of e-mail accounts) increase the number of goods (computer systems, e-mail links) that the network offers to all participants. Thus, the value an individual assigns to a specific system increases with the number of such systems sold. In this sense, both concepts are closely related.

---

[7]   As there is no point of owning an e-mail account if there is no one an e-mail can be sent to and received from, the "stand-alone value" in such communication networks typically goes even down to zero.

[8]   Pioneering works in this area are Chou and Shy (1990, 1993) and Church and Gandal (1992, 1993).

[9]   The concept of indirect network effects is also referred to as *hardware / software paradigm* (see, *e.g.*, Church and Gandal 1992 or Katz and Shapiro 1994)

[10]   See, *e.g.*, Church and Gandal (1992, 1993, 1996).

[11]   Precisely, this result can be obtained if one assumes that software is provided by monopolistic competitors (including free entry) and each firm in the market offers one variant of software (see Chou and Shy 1990).

A further source of network effects mentioned in the literature is decreasing marginal costs in production. According to this concept consumers benefit from large sales because prices are expected to be lower. This approach is conceptually different from the concepts of direct and indirect network effects as defined here. These two concepts focus on the gross value of the goods (e-mail account, hardware), while the concept of economies of scale in production focuses on the price of goods.[12] We concentrate our analysis on the above-derived concepts of direct and indirect network effects as pure adoption externalities. In fact, cases where consumers' *net benefits* decrease if networks enlarge are not excluded.

Potential relevance of network effects

One reason for the relevance of network effects derives from the ever-increasing significance of Information Technology (IT) in society today. Since ITs are based on communication (*i.e.* data transfer), standards are essential for the functioning and interworking of IT systems. For example, the decision on the tools and formats used to design an Internet-site almost exclusively depends on the coverage of browsers and plug-ins among Internet users. In addition, there are many virtual networks that play an important economic role. For instance, consumers prefer widely used software as the possibilities for easy data exchange increase. Further examples alluded to in the literature where both direct and indirect network effects may be involved, are measurement systems (metric vs. Anglo-Saxon imperial) and systems to define and classify goods (*e.g.* RAL for colors or IPE for steel carrier). Other typical products include ranges, photo cameras (because of the need compatible of film), VCR systems[13], Automatic Teller Machine (ATM) services (as a particular service with many customers may possess a higher density of ATMs), television (over-the-air and cable), electricity networks, retail dealer networks, the list goes on and on.

Also, it is claimed that financial markets exhibit network effects. Economides (1993) argues that market liquidity increases when the number of traders increases. More precisely, a growing number of traders at a given stock exchange augments the benefits of all its traders because it is easier to find a counter-offer. Related to this line of reasoning, Berndt (2000) shows that a firm's ownership structure exhibits network effects. The greater the number of participants in a financial system who choose either "dispersed" or "concentrated" ownership, the more effective the respective institutions become in reducing agency costs. For example, with dispersed ownership, the take-over market is necessary to discipline management. The functioning of this mechanism, however, depends on the liquidity of stock markets. Liquidity, in

---

[12]   See, *e.g.*, Holler (1996), Liebowitz and Margolis (1994) and Katz and Shapiro (1986) for a deeper discussion.

turn, increases with the number of firms that are held in dispersed ownership. Applying the metaphor of actual networks, the larger the number of firms with dispersed ownership, the larger the number of "potential take-over-links" between them.

In a series of articles, Kahan and Klausner (1995,1996,1997) suggest that one firm's corporate contract can positively influence the value of other firms' contracts. Indirect network effects with corporate charter terms (interpreted as hardware) are claimed to emanate from several sources. Widely used corporate charter terms reduce present and expected uncertainty associated with that rule. The more firms adopt a particular term, the more likely it is the term will be litigated, and therefore the more likely that future judicial interpretations (interpreted as software) will be provided. In addition, "common practices" that comply with the term will develop among firms that implement the term. Furthermore, economies of scale are, to some extent, involved in the production of legal services. Thus, superior legal advice will be available to a firm which must interpret or implement that term. Moreover, how familiar investors are with a particular term may have an effect on the marketability of the firm's securities, as the investors can assess the firm's value more easily. Investors' familiarity with a particular charter term, in turn, tends to increase with the number of firms using that term.

Many scholars stress that the phenomenon of network effects is important within the context of social and legal norms. Since these are supposed to govern social interactions, it is not surprising that the adoption of many norms involves adoption externalities.[14] "The people in Europe and their governments, for example, benefit greatly in politics, business and social life by commonly acknowledging 'western democratic values'. These agreed standards avoid complicating disputations and their possible conflicts and contracting on basic principles in everyday transactions" (Adams 1996, p. 370). Another important example is the use of languages and letters. Obviously, the value an individual assigns to learning a specific language and system of letters increases with the number of people who are able to communicate with it.[15] See Shy (2001), chapter 10 and 11, for more examples on social norms and network effects.

Finally, it should be pointed out that network effects could turn into negative external effects as well. When too many individuals join a network, the quality of transmission (-links) may deteriorate and thus decrease consumers' evaluation of network participation. Such ca-

---

[13] VCR=Video Cassette Recorder.

[14] Within the context of legal norms, often the "adoption" is not done on the individual level. Rather, adoption externalities may concern lawmakers' decisions on which laws to "adopt". See, e.g., Berndt (2000) for a study on lawmakers' choice under network externalities in financial markets.

[15] See, e.g., Holler and Wickström (1999), Posner and Rasmusen (1999), Störig (1992), Blankart and Knieps (1993).

pacity problems apply mainly to physical networks and are, moreover, likely to be rather short-term. MacKie-Mason and Varian (1996, 1995), for example, analyze this issue with application to peak-load pricing in the Internet. Of course, adoption externalities can even be negative from the start or if not negative from the beginning may subsequently become negative even for a small number of adopters. Examples include goods like exotic cars or designer furniture whose users derive benefits from exclusivity.

2.2   Basic Problems in Competitive Network Markets

Too small networks

Due to external adoption effects, the price mechanism might fail to optimally coordinate actions. In principle – no matter whether they are direct or indirect – network effects change the demand function from $p(n)$ in $p(n; n^e)$, with $\delta p/\delta n^e \geq 0$, where n^e represents the number of expected sales.[16] Thus, the expected network size is similar to a quality parameter. With a stand-alone value of zero, a typical fulfilled expectation demand function $p(n; n)$ may equal the one shown in Figure 1.

In Figure 1, each $p(n; n^e_i)$ represents the demand for given expected sales n^e_i. The points E_i, where n^e_i equals n_i, give fulfilled expectations. The locus of all E_i plus the vertical axis yield the fulfilled expectation demand curve, $p(n; n)$.[17] If network effects are sufficiently strong the fulfilled expectation demand increases for small n as shown in Figure 1.[18,19]

Even if agents manage to coordinate in the Pareto-optimal equilibrium, sales are too low from a societal point of view. If an agent joins a network, he confers benefits also on other participants. Consequently, if network access is priced with marginal costs, private incentives for joining the network are too low. Moreover, it is not clear that agents coordinate in the Pareto-optimal equilibrium in the first place.[20] Figure 1 exhibits that network markets are very

---

[16]   See also Holler (1996) for an overview of how network effects can be modeled.

[17]   With a stand-alone value of zero, zero sales is always an equilibrium.

[18]   See Economides (1996) and Economides and Himmelberg (1995) for more details.

[19]   In cases of direct network effects, if the usage of links (communication, exchange of data etc.) is charged, these prices would enter the demand function, too. Similarly would do prices for software in cases of indirect network effects. In the remainder of the study, it is abstracted from such effects. We either assume that costs of links usage are unaffected by the networks' sizes and providers of networks charge marginal cost prices for use of links or we suppose that firms do not control links. For most examples mentioned in section 2.1, the latter applies. For instance, suppliers of software do not charge users for exchange of data and producers of VCRs cannot control exchange of videocassettes. Obviously, this also applies for interactions under particular contract terms and social norms.

[20]   It seems, however, to be a standard assumption in the network effects literature that players manage to coordinate in the Pareto-optimal equilibrium (see, *e.g.*, Fudenberg and Tirole 2000, Economides 1996 and Katz and Shapiro 1992, 1986; one exception is Farrell and Katz 1998).

sensitive to expectations as there may be more than one market equilibrium (the marginal cost curve (c') and the fulfilled expectation demand curve intersect three times). Hence, with a given price, a coordination problem may exist, related to a "simple coordination game" as displayed in Figure 2.

Figure 1

Fulfilled Expectation Demand Function

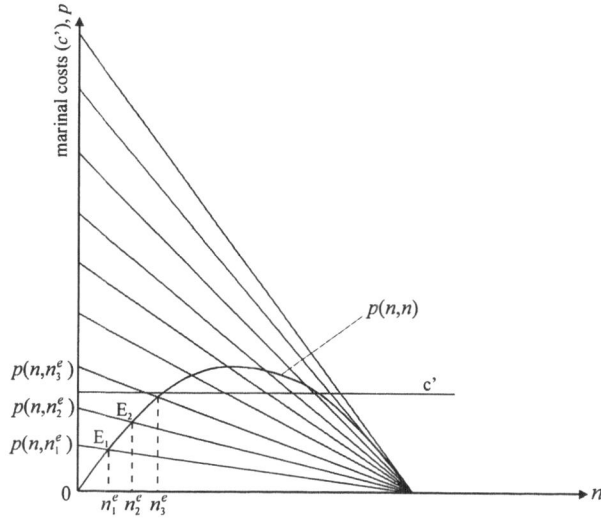

Source: Economides and Himmelberg (1995), p. 7.

Figure 2

Game-matrix "Join a Network"

Agent 2

		Join	Not join
Agent 1	Join	(3,3)	(0,0)
	Not join	(0,0)	(1,1)

Due to the existence of multiple equilibria, *i.e.* (Join, Join), (Not join, Not join) and an equilibrium where players choose mixed strategies[21], it might occur that only one or no agent joins the network. Restricting attention to pure strategies, one could expect that this kind of coordination problem would be aggravated, the more the payoffs of the two pure strategy equilibria are equalized.

The assumption that players move simultaneously might represent an ex ante situation, where it is too costly to wait for the choices of other players. For instance, this might be the case if there is a time lag between an investment (unobservable by other players) and its return (which depends on both players' investment choices). In a setting where players move sequentially the problem may diminish. Subgame-perfection suggests that players coordinate themselves such that both end up joining the network.[22]

In a similar setting where, however, the order of players' moves is *not* predefined, it may be a dominant strategy to wait for the choice of other players. In this case, no player may join the network because expected demand (n^e) in each time is zero. Players face a problem of collective action, possibly resulting in no network at all being established. This applies if networks have to pass some *critical mass* in order to materialize in the first place.

Too much or too little variety

Most parts of the study focus on problems that arise in rival networks. On the one hand, if agents prefer different (network-) goods then they benefit from *variety among networks*, since consumed goods are better tailored to different tastes. On the other hand, they prefer large and thus a rather small number of networks in order to exploit network benefits. Thus, compared to markets without network effects, network markets tend to have a lower level of variety, which is desirable from a social point of view too. When agents choose a product they take into account *their* benefits from (expected) network size, thus potentially reducing the use of different network variants. Therefore, if network effects are sufficiently strong, such markets may even be characterized by "tipping", *i.e.* only one variant (the "*standard*") survives.[23]

All the same, there remains a tradeoff between networks' size and networks' variety. It is, however, not guaranteed that agents' optimal solution to this tradeoff implements the socially optimal outcome. Agents contemplating joining a network do not take into account the part of

---

[21]   Both players assign probability of ¼ to play "join" and ¾ to play "not join".

[22]   Nevertheless, also in the setting with simultaneous moves, there exist equilibrium refinement methods: Harsanyi and Selten's payoff-dominance criterion suggests that both players choose "join" (see Harsanyi and Selten 1988 for this criterion).

[23]   See, *e.g.*, Shapiro and Varian (1999a, 1999b).

network benefits that is conferred on other agents. Hence, they might decide to join the "wrong" network from a social point of view, resulting in too many or too few networks.

In a static setting where agents move simultaneously, agents may face a coordination problem similar to the one described above. With strong network effects and different preferences of users (with regard to the goods given equal network size), the situation may resemble a Battle of the Sexes game as shown in Figure 3.

Figure 3

Game-matrix "Choose a Network"

Agent 2

		Network 1	Network 2
Agent 1	Network 1	(3,1)	(0,0)
	Network 2	(0,0)	(1,3)

Intuitively, the likelihood that agents manage to coordinate themselves is even lower than in the "simple coordination game" displayed in Figure 2 because the state in which they prefer to coordinate themselves differs among them.[24]

Again, in a dynamic setting with sequential moves the coordination problem may diminish because subgame-perfection suggests that both players end up joining the network that the first moving player prefers. In a setting without a predefined order of moves, agents might again suffer from collective action problems. For instance, even if every agent was better off switching from one network to another – provided a sufficient number of agents actually does so – they might stay in their original network. If agents can follow the first movers without relevant losses, it may be a dominant strategy to wait and see what others do. Analogous to the above case of emerging networks, if agents are not able to coordinate their actions, the expected number of agents that switch is zero. Obviously, this is sub-optimal, as agents from the other network would benefit from that switch. A similar case could arise if agents are situated in one network but prefer to establish a variety state.

Excess inertia and excess momentum

Agents might undesirably stick to an established inferior network technology (excess inertia) or agents might undesirably switch to a new emerging one (excess momentum). The rationales behind this are closely related to those discussed in rivaling and emerging network. With simultaneous moves, the nature of the problem may be equivalent to the coordination games shown in Figure 2 and Figure 3, if we rename players' strategies with "Stay" and "Switch", respectively.

Farrell and Saloner (1985, 1986a) model the problem by means of a sequential game and show that both "excess inertia" and "excess momentum" can occur. With excess inertia, agents stay with the inferior network (too long) although transition to a new superior one is socially desirable; with excess momentum, the opposite is true. The latter may happen if the losses accrued to "stranded" users are greater than the gains to those who benefit from transition. For a simple example take a payoff structure similar to the Battle of the Sexes game in Figure 3, however here assume asymmetric payoffs and sequential instead of simultaneous moves, as shown in Figure 4:

Figure 4
Game trees: excess inertia and excess momentum

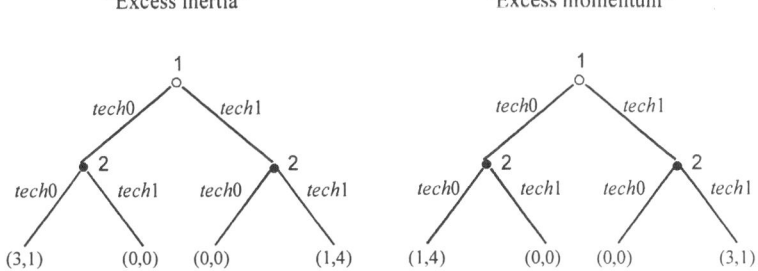

The action "*tech0*" represents "staying" with the established network-technology while "*tech1*" stands for "switching" to the new one. The unique subgame-perfect equilibrium in the game on the left hand side (*lhs*) and the one on the right hand side (*rhs*) is (*tech0, tech0*) and (*tech1, tech1*), respectively. Assuming a utilitarian social welfare function, the outcome of the game on the *lhs* represents excess inertia because joint payoffs with transition to the new network technology would be larger. In contrast, in the game on the *rhs*, excess momentum occurs. Utilitarian welfare would be higher if both agents stayed with the established technology.

---

[24]   Again, Harsanyi and Selten (1988) offer a criterion for equilibrium selection in such games.

In a setting where every agent could (profitably) switch at all times, however, transition from an established network to a new superior one may be hampered by collective action problems similar to those discussed above. Agents may have an incentive to hesitate before switching in order to not relinquish network benefits. Consequently, if agents cannot coordinate such a switch, nobody wants to go first, and so all agents might stay with the inferior network.[25,26]

To conclude, due to adoption externalities, the optimal performance of competitive network markets is threatened by several effects: agents who contemplate joining a network do not take account of the benefits conferred on other consumers within the network. In addition, there might be coordination as well as collective action problems. These effects might lead to networks being too small, result in too much / little network variety, as well as technological excess momentum and excess inertia.

2.3   Compatibility, Standardization, Harmonization and Integration

The notion of compatibility

As demonstrated in section 2.1, participants' of a network derive benefits from having links to other nodes (participants), *i.e.* from combining complementary components. Whether such links actually exists or not, is a matter of *compatibility*. "We have pointed out that links on a network are potentially complementary, but it is compatibility that makes complementarity actual" (Economides 1996, p. 676). For illustration, consider fixed-line and mobile telephony. Since links exist between participants of both "sub-networks", *i.e.* one can call someone on his mobile phone from a fixed line phone (and vice versa), the associated network benefits extend over *both* sub-networks. In contrast, if calls were only possible within each sub-network (*i.e.* sub-networks are incompatible), each sub-network would then provide network benefits only according to the number of links within it. Likewise, networks of users of word processing software are limited to those who use compatible software. So, there exists a network of *Microsoft's WORD* users and one composed of *WordPerfect* users.

---

[25]   An early contribution on this issue is David (1985) where the well-known example of the QWERTY keyboard is introduced. Other pioneering works within this area are Farrell and Saloner (1986a) and Katz and Shapiro (1986).

[26]   In this sense, a technological "revolution" possesses features similar to a "political" revolution where the first people who act against established government incur the highest risk of getting punished. The more people have already begun to revolt the less dangerous it becomes to join the revolution.

How to exploit network effects

Suppose there are two rival incompatible networks. What can players (firms, users, policy makers, etc.) do in order to exploit network benefits? Tautologically, one way to do so is to make the networks (more) compatible. This process is commonly referred to as *standardiza-tion*.[27] That is, those designing the goods, *i.e.* producers, lawmakers, etc., modify their goods (products, norms, etc.) so that they are compatible. For example, producers decide how to design their telephone networks and computer programs with respect to their ability to make them work with substitute ones. Similarly, lawmakers may stipulate accounting rules, which are easy to read and interpret for firms from other countries that adopt different accounting rules.

Figure 5

How to exploit network effects

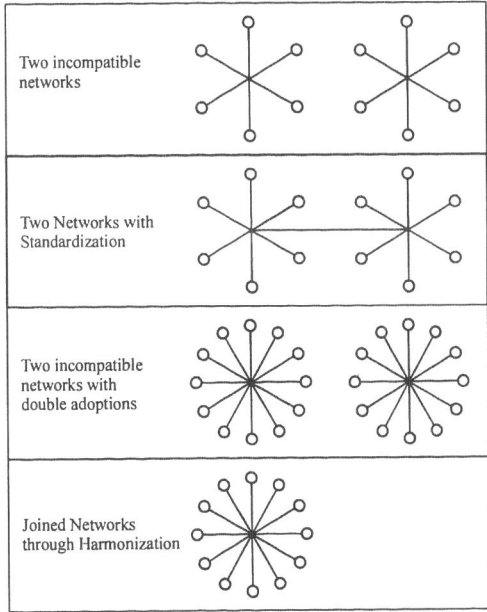

---

27   See, *e.g.*, Farrell and Saloner (1987).

If suppliers of goods fail to provide for compatibility, users may themselves produce compatibility through buying *adapters* (if available), such as adapters for plugs or software that allows translating *WordPerfect* documents into the *Microsoft's WORD* format. Often, however, incompatibility lies in the nature of the goods themselves or a converter is too costly, *i.e.* these goods are "inherently" incompatible. Similar to buying adapters, users may, in some of these cases, help themselves and make *"double-adoptions"*. For example, many people speak more than one language.[28] In addition, such double-adoptions are typical for computer software where many users have installed several functionally equivalent browsers like *Internet Explorer* and *Netscape* at the same time.

Double adoptions are not only limited by the costs of network goods. Often, limitation of networks' sizes is given by natural "mutual exclusivity". That is, irrespective of the costs of the network goods, one has to choose a single one. For example, firms can only have one kind of ownership structure, contracts can arrange a subject with only one kind of term, etc. In such a situation, the only remaining way to exploit network benefits is *harmonization, i.e.* all agents belong to the same network. For instance, people in the world would all learn English, adopt the same type of software and all firms would have the same ownership structure, etc.

See Figure 5 for a summary. Each outlined node represents a user of a network good. Depending on the particular network good, links stand for possibilities of communication or data exchange or for low transaction costs, etc. A similar figure applies to systems with indirect network effects.

Standardization vs. harmonization

With perfect and costless standardization, the potential problem of too many or too few networks as well as the problem of excess inertia and excess momentum (see section 2.2) disappears. Obviously, harmonization may, at most, only reduce these problems. While harmonization causes users to better exploit network effects, it also imposes losses of network-variety on users.

Note, however, that in most cases, standardization involves some harmonization as well. In order to achieve standardization, networks often have to be aligned with each other, *i.e.* differences between them are reduced. As Farrell and Saloner (1992) point out, standardization includes doing certain key things in a uniform way.[29] This is most obvious for goods that consist of different components, where standardization requires harmonization of interface com-

---

[28]   There do exist converters in the form of human or electronic translators. They are rather imperfect and
       costly, though.
[29]   See Farrell and Saloner (1992), p. 9.

ponents, such as the plugs of computer devices. Thus, even with standardization, there might be a tradeoff between network benefits and networks' variety.

Moreover, standardization itself can be imperfect. Rather than being a matter of "whether or not", compatibility is often a matter of degree.[30] In such a case, the optimal solution is a tradeoff between the degree of standardization and variety among networks.[31] For example, one often observes that software programs can read data sets from other programs. However, in most cases transformation is imperfect and it takes additional time to further prepare data sets before using them. Further equalizing the functions of the programs would probably increase the compatibility between them while it would sacrifice some variety among them. For another example consider these two strategies. The first one entails learning English as your native language and German as your second language. The second strategy is to learn German first and English as a second language. These strategies are probably imperfectly compatible because conversation between two people would be smoother if both would have adopted the same language as the native one. On the other hand, the degree of compatibility of these strategies is probably higher the more similar the languages are in the first place.

Integration of networks

An additional major factor within the tradeoff between harmonization (or standardization) and variety is the degree of integration among different networks. For illustration consider two (incompatible) communication networks, which are differentiated by the language in use. Imagine that, for exogenous reasons, users only wish to talk to people from the same network. Obviously, the optimal solution to any tradeoff between network benefits and networks' variety entails 100% percent variety. Network benefits would remain constant even if both networks were perfectly standardized. How do network benefits behave when two incompatible networks become more integrated? Given that the intensity of links' usage (*i.e.* frequency of conversations) remains constant, network benefits *decrease* because people can derive benefits from complementarity only if they communicate with someone from their own network. Thus, the more integrated networks are, the larger the benefits from standardization and/or harmonization.

This idea of network integration directly leads us to the problem of international harmonization, which is the subject of section 3. Different (incompatible) national standards form dif-

---

[30] Formally, the level of compatibility can, for example, be expressed by weighted network benefits. For example, while participants of network A receive network benefits from network A by weight 1, the network benefits from network A conferred to network B is weighted with a value between 0 (incompatibility) and 1 (perfect compatibility). See Holler (1996) for literature examples.

[31] Blankart and Knieps (1993) is an early contribution within this line of thinking.

ferent national networks. The more advanced integration (*i.e.* globalization) is among countries, the lower are the network benefits generated by national standards and the more desirable is standardization and/or harmonization. Adams argues similarly: "... an arbitrary combination of social norms from different cultures has the disadvantage that people would lose the benefits of compatibility produced by norms of one culture" (Adams 1996, p. 370). These arguments indicate that the degree of globalization among countries is a crucial parameter when analyzing the evolution of international standards.

Quality standards and network effects

Although their nature is often claimed to be different, note that so-called "quality-standards" often possess underlying characteristics of compatibility standards. To see why, draw an analogy to the example of languages as networks. Quality standards are often just "definitions" or names of products, which stand for the products' composition of ingredients or relevant characteristics. As with languages in general, producers and consumers benefit from using the same "names" for equivalent products because transaction costs are reduced. To learn and adopt these names corresponds to the act of network access, which, in turn, produces links for low-cost interactions between producers and consumers. The more producers and consumers join the network the more beneficial it is to adopt such a name. Examples include quality standards for food, or steel and concrete within the civil construction industry.

Of course, due to the unobservable quality of experience and credence goods at the time of purchase, further problems may arise within the context of quality standards for those goods. However, if moral hazard and adverse selection problems are eliminated, *e.g.* through signaling devices via price, advertising, warranties, etc. (see Schäfer 2001, Davis 1992, Milgrom and Roberts 1986) and/or reputation mechanisms, such quality standards reveal themselves in effect to be pure compatibility standards.

2.4   Commitments

It has been demonstrated in sections 2.2 and 2.3 that outcomes in network markets heavily depend on the specific modeling, especially on how players' moves are predefined. For example, if all users are completely informed and can switch between networks without costs at any time, collective action problems are likely to occur. In contrast, with sequential moves, we can expect that users manage to coordinate in a Pareto optimal equilibrium (if it exists). As they determine the appropriate way of modeling, switching costs or "commitments" may be crucial for the analysis of network industries. This supposition will be verified in section 3.7.2 and 5. For future reference, we briefly explain here why such commitments may evolve.

A dynamic analysis of network markets entails that agents might wish to switch from one network to another. First, their initial purchase of a network good might ex post prove unfavorable, *e.g.* their expectations with regard to choices of other agents have not been fulfilled. Second, the environment might have changed over time. For instance, a new "superior" network good may have emerged. Third, agents may wish to possess a technology compatible to their actual opponents, who, in turn, might change frequently.

Switching between networks involves costs if the investments (the purchases of the durable goods) are partly or wholly sunk.[32] Without any sunk costs, an agent switches to another network if it yields higher net-benefits to him, because he can resell his "old" network good and buy the "new" one. With sunk costs, however, a user may stick with his once chosen network, even if another network's net benefits are higher. Thus, sunk costs produce some commitment. See Table 1 for a summary.

Table 1
Network Goods and Commitments

Network good is free or not durable	Network good is costly and durable	
	Not sunk	*(Partly) sunk*
No Commitment		Commitment

There are several possible reasons for sunk costs. For example:

- Agents may have to incur price-drops due to imperfection of the market for used goods, which might, in turn, be caused by moral hazard or adverse selection in that market. Both phenomena can prevail if buyers in the market for used assets face some quality uncertainty. We encounter moral hazard if users treat their asset worse than efficiency would suggest (unnoticeable by a potential buyer). That is to say, since users do not use the asset over its entire life, they have an incentive not to incur efficient expenses to enhance the asset's life and other quality parameters (*e.g.* expenses for maintenance). Second, as a result of adverse selection, the quality of goods offered in the second-hand market tends to be lower than the quality of the entire set of these goods in use. When

---

[32] Of course, there are other sources of switching costs. For an overview, especially for those producing brand loyalty, see Klemperer (1995), pp. 517-518.

potential buyers cannot observe the goods' quality, invariably owners of goods whose quality has proven to be low have incentives to offer their goods (see Akerlof 1970).

- Expenses for the goods include some transaction costs such as search costs, costs of ordering, costs of installing, as well as costs of specific training, etc. (see, *e.g.*, Matutes and Regibeau (1996), p. 188).

- Similarly, the good itself is not marketable or "remarketable". This applies to the choice of contract terms, social norms and similar "goods". Correspondingly, software vendors often preclude their clients from reselling the software.

2.5 Compatibility and Competition

Most parts of the study are based on *inherently* incompatible goods or technologies. In section 2.1, we have demonstrated that such network markets tend to have a lower degree of variety than "regular" markets. Sufficiently strong network effects may even urge competitive markets to tip to one out of several incompatible variants, *i.e.* market forces themselves establish harmonization, which, of course, then implies that all used goods are compatible with each other.

As described in section 2.3, in many cases, the goods themselves are not inherently incompatible, rather it is those who design them that actually determine the levels of compatibility.[33] As demonstrated in the previous sections such decisions may affect consumers' benefits and thus market performance. Katz and Shapiro (1986) claim that "when network externalities are large, the choice whether to make the products or technologies compatible is one of the most important dimensions of industry performance."[34] Although most parts of our study deal with inherently incompatible goods, we want to briefly analyze incentives of firms to make their goods compatible. This analysis shows that firms often have an incentive to keep incompatibility even if social welfare with compatibility is higher. Thus, even those parts of the study that are based on the assumption of inherently incompatible goods may – to some degree and extent – apply to potentially compatible goods as well.

---

[33] If we considered goods where firms can also price the usage of actual or virtual links, then the degree of compatibility would not only depend on the design of the goods but also on the prices that firms charge for use of links between different networks. For example, mobile telephony providers often charge a higher price for calls to another network than for calls within the network. As noted already in section 2.2, we abstract from these effects; with most "virtual" network goods, as the examples in the previous sections reveal, firms cannot charge for the usage of links.

[34] Katz and Shapiro (1986), p. 154.

Compatibility incentives with "perfectly" differentiated goods

With perfectly differentiated goods, production of compatibility is similar to the provision of public goods. For illustration purposes, assume that there are two types of consumers, A and B. A-types only buy A-goods and B-types only buy B-goods. Normalize demand for each kind of good to 1. Consumers' willingness to pay depends on whether goods A and B are compatible with each other. If the two kinds of goods are compatible with each other, willingness to pay is P_A^C and P_B^C, otherwise it equals P_A^I and P_B^I, respectively. Due to network effects $P_A^C > P_A^I$ and $P_B^C > P_B^I$.

To illustrate these assumptions, imagine there are two groups of consumers that differentiated by their members' (strong) preferences; for instance, the groups may represent actual countries. Alternatively, populations stand for industries. For example, A-types are members of the advertising community who strongly prefer Apple computer systems and B-types are more technical industries whose members prefer Windows systems. Obviously the members of both industries prefer compatibility of systems, as advertising firms want to exchange data with their clients from the technical industry (and vice versa).

Constant marginal costs are c. Making goods compatible increases marginal costs by Δ. Coming back to the metaphor of actual networks, production of compatibility equals the construction of links to the other network (see sections 2.1 and 2.3). It is assumed that such links can be used in both directions, *i.e.* both A- and B-consumers' willingness to pay increases up to P_A^C and P_B^C, respectively, if one firm makes goods compatible. Note, however, that the willingness to pay of both kinds of consumers increases even if the link between networks could be used in only one direction (say from A to B). Returning to the previous example of computer systems, B-consumers would benefit also if A-consumers could receive data from them. Consequently, once a firm has established links to the other network, it may have strong incentives to allow it be used in both directions, as this also increases the willingness to pay of its clients. Thus, although the strength of the conclusions may be slightly diminished, the setting essentially applies for the case of "one-way" compatibility as well.

Suppose first, each kind of good is offered by one firm, firm A and firm B, respectively. Thus, $P_A^C - P_A^I$ (and $P_B^C - P_B^I$) represents additional returns of firm A (firm B) when goods A and B are compatible.

If

(C1) $\qquad P_A^C - P_A^I + P_B^C - P_B^I > \Delta > \max[P_A^C - P_A^I, P_B^C - P_B^I]$

it is a dominant strategy for both firms not to incur Δ although it is socially desirable that one firm does so.[35] Through compatibility, a firm's profits increase by $P_A{}^C - P_A{}^I < \Delta$ and $P_B{}^C - P_B{}^I < \Delta$, respectively, while total profits increase by $P_A{}^C - P_A{}^I + P_B{}^C - P_B{}^I > \Delta$.

The problem is reduced – however it does not disappear – if Δ is smaller. If

(C2) $$0 < \Delta < \min[P_A{}^C - P_A{}^I, P_B{}^C - P_B{}^I]$$

firms face a situation similar to a "chicken game". See Figure 6 where profits with incompatibility are normalized to zero. The game possesses three equilibria. In two of them, one firm incurs Δ while the other firm does not. The third equilibrium is in mixed strategies, *i.e.* both firms incur Δ with some positive probability. The outcome of such a game is uncertain. Although compatibility is socially desirable, firms and consumers may end up with incompatibility.[36] Moreover, both firms may incur Δ although it is socially better if only one firm were to do so.[37]

Figure 6

Compatibility decision as chicken game

Firm B

		Comp	Incomp
Firm A	*Comp*	$P_A{}^C - P_A{}^I - \Delta$; $P_B{}^C - P_B{}^I - \Delta$	$P_A{}^C - P_A{}^I - \Delta$; $P_B{}^C - P_B{}^I$
	Incomp	$P_A{}^C - P_A{}^I$; $P_B{}^C - P_B{}^I - \Delta$	0; 0

The game's outcome may be socially superior if Δ is larger such that one firm's profits increase by more than Δ while the other firm's profits increase by less. For example, let:

(C3) $$P_B{}^C - P_B{}^I < \Delta < P_A{}^C - P_A{}^I$$

---

[35] A similar logic can also be found in Berg (1989).

[36] In mixed strategy equilibrium, the probability that firms assign to "*Comp*" even decreases with shrinking Δ although the social desirability increases.

[37] We will see that such a problem does not only apply to producers' compatibility choice. A similar situation may also arise for users' compatibility choice (buying adapters or making double-adoptions). See our discussion in section 3.8 for details.

If condition (C3) is satisfied firm A establishes compatibility. This situation might resemble standardization by a dominant firm as it is observed in some industries.

All three above situations typically occur in a public good context. Of course, if condition (C1) or (C2) is satisfied, the two firms have an incentive to conclude a contract that deals with their compatibility choices and the division of Δ. A necessary condition for such a contract to occur is enforceability. It depends on the specific industry whether this kind of contract is enforceable. Probably, there are cases where, due to complexity of the goods, firms would face difficulties to prove that the other firm has breached the contract, especially when the minimizing of total compatibility costs requires the redesigning of *both* goods.

If both kinds of goods were offered by a large number of firms, transaction costs may become prohibitively large and deter firms from contracting. Free-riding incentives, however, remain similar if both goods are supplied competitively. To see why, assume for simplicity that each kind of good is offered by two firms (firms A_1, A_2, B_1 and B_2), which behave like Bertrand-competitors.

Without loss of generality, consider A-firms first. Let condition (C1) hold, *i.e.* assume it is a dominant strategy for both monopolists to maintain incompatibility. Given firm A_2 keeps incompatibility, then if firm A_1 renders its goods compatible to B-goods it captures the entire market for A-goods. The most firm A_1 can earn is $P_A{}^C - P_A{}^I - \Delta$.[38] This equals the extra profits of the A-monopolist because firm A_1 is "disciplined" by its competitor, firm A_2. That is, a share of the extra rent generated by firm A_1's compatibility decision accrues not only to B-consumers but also to A-consumers. Since the same rational applies to firm A_2 as well as to each B-firm, incompatibility remains.

Now let condition (C2) hold. Recall that this condition produces a chicken game in the monopoly case. To analyze Bertrand competitors imagine that B-firms choose first. Afterwards A-firms simultaneously decide whether to make their goods compatible to B-goods. Payoffs may look like those given in the matrices in Figure 7.

---

[38]   In Bertrand competition, the price of incompatible A-goods equals the marginal costs, c. Thus, firm A_1 must offer A-consumers a rent at least of $P_A{}^I - c$, *i.e.* it sets a price equal to $P_A{}^C - P_A{}^I + c$. Consequently, firm A_1's extra returns (before deducting Δ) amount to $P_A{}^C - P_A{}^I$. The incentives to establish compatibility might be larger if firm A_1 expects firms A_2 to do so. However, both firms would be worse off with compatibility. Thus we assume that both refrain from doing so.

Figure 7
Payoffs of Bertrand competitors

a) No B-firm establishes compatibility

Firm A_2

		Comp	Incomp
Firm A_1	Comp	0; 0	$P_A^C - P_A^I - \Delta;\ 0$
	Incomp	$0;\ P_A^C - P_A^I - \Delta$	0; 0

b) One or both B-firms establish compatibility

Firm A_2

		Comp	Incomp
Firm A_1	Comp	0; 0	0; Δ
	Incomp	Δ; 0	0; 0

The payoffs in Figure 7a apply to A-firms if B-firms have kept incompatibility. Both A-firms would render their products compatible. If firm A_1 makes its goods compatible while firm A_2 leaves incompatibility, firm A_1 captures the entire market and receives $P_A^C - P_A^I - \Delta$ > 0. Since firm A_2 makes zero sales, its profits are zero. If both A-firms make goods compatible, competition drives profits down to zero, which coincides with the profits when both firms leave incompatibility. Hence, leaving incompatibility is weakly dominated. Trembling hand perfection suggests that firms make goods compatible. (Admittedly, the payoffs in Figure 7a are highly stylized. Note, however, that prediction becomes even stronger if payoffs are more "realistic". Firm A_2's payoff with (*Comp, Incomp*) is probably less than zero, similarly firm A_1's with (*Incomp, Comp*). In addition, one might suppose that firms' payoffs with (*Comp, Comp*) are greater than with (*Incomp, Incomp*).)

Figure 7b applies to the case where one or both B-firms have established compatibility. Both A-firms keep incompatibility. Every consumed B-good is compatible to the A-goods, even if only one B-firm makes its goods compatible and A-firms choose incompatibility (as that B-firm, which chooses compatibility captures the entire market for B-goods). Thus, for A-firms, it does not make sense to incur Δ because consumers' willingness to pay is unaffected. If firm A_1 chooses "*Incomp*" while A_2 goes for "*Comp*", firm A_1 has a cost advantage and satisfied all of the demand, and it earns profits of Δ.[39] Were both firms to render their goods compatible, then competition drives profits down to zero. The unique trembling hand perfect equilibrium is (*Inc, Inc*). Consequently, if they expect that one or two B-firms establish compatibility, then both A-firms maintain incompatibility. (Again, the payoffs are highly stylized and prediction probably becomes even stronger if "more realistic" payoffs are given. Firm A_1's payoff with (*Comp, Incomp*) is probably less than zero, similarly firm A_2's with (*Incomp, Comp*). In addition, we might suppose that firms' payoffs with (*Incomp, Incomp*) are greater than with (*Comp, Comp*).)

Analogous payoffs and results are derived for B-firms if A-firms choose first.

For the game where all firms simultaneously decide whether to make goods compatible, we may reduce each group of firms, A and B, to a single player. Deriving payoffs from Figure 7, the game is similar to a chicken game and consequently possesses two pure strategy equilibria, (*Comp, Inc*) and (*Inc, Comp*), and one equilibrium in mixed strategies. Thus, the chicken game structure may remain even in a competitive environment, therefore, whether compatibility occurs or not is uncertain. If an analogous condition to (C3) holds, we can expect that both A-firms establish compatibility, since this situation is equivalent to A-firms reacting to B-firms having kept incompatibility.

To summarize, independent from whether the goods are supplied competitively or by a single firm for each kind of good, market forces may not induce compatibility even if it is socially desirable. A firm's incentives to establish compatibility are too low because a share of the generated rent confers on those firms and consumers that offer and consume, respectively, different but (then) compatible products. Incentives for free-riding may further exist because firms speculate that those from the other market bear the costs of compatibility. Were goods offered by monopolists, firm's profits are maximized if the other firm establishes compatibility. In the end, the incentives of competitive firms are similar. A competitive firm is more worried about its relative position to its competitors. If firms from the other market establish compatibility it risks having higher costs than its competitor(s). On the other hand, if firms

---

[39] Cost advantage is Δ. Thus, firm A_1 can offer its good with a mark-up close to Δ and still capture the entire market.

from the other market keep incompatibility, a domestic firm's incentive to render its goods compatible is strong: if it leaves its goods incompatible it risks offering an inferior product. We have shown that in both the monopolistic and the competitive case, firms are potentially exposed to a chicken game. In this game, incompatibility may be the outcome even if compatibility is highly desirable. In our specific setting, only when it is a dominant strategy for one firm (or one group of firms) to keep incompatibility (*i.e.* (C3) holds), will desirable compatibility result with certainty.

Compatibility incentives with "imperfectly" differentiated goods

Now assume that consumers' preferences are less heterogeneous than those in the above model, such that all firms compete for all buyers. We shall present two effects that may arise in such settings. One is in favor of compatibility; the other one is in favor of incompatibility.

Shy (2001) analyzes a duopoly that sells two differentiated brands to heterogeneous consumers. He argues that firms which compete on price may favor compatibility because competition in an incompatibility regime is stronger. To see why, remember that if both brands are compatible, then network benefits enjoyed by users do not depend on the number of users of the specific brands, but on the *entire* number of users of the good. In contrast, if brands are incompatible, each brand provides network benefits only according to the number of its own users. Consequently, in the incompatibility regime, each firm's incentives to undercut its rival are stronger because any additionally attracted user makes its brand even more attractive. In contrast, price competition is relaxed when brands are compatible, "since under compatibility firms' network size becomes irrelevant to consumers' choice of which brand to buy" (Shy 2001, p. 31). Thus, both firms are likely to produce compatibility in such a setting. Moreover, by reducing competition, firms may appropriate an over-proportional share of the generated (social) rent through compatibility. Therefore, if production of compatibility is costly, firms' incentives to establish compatibility might even be too strong from a social point of view.

Suppose now that network effects are very strong. As argued in section 2.2, such markets are prone to tip to one out of several incompatible substitute goods. Thus, as Katz and Shapiro (1994, 1986) argue, there are strong winners and strong losers under incompatibility. Consequently, a firm that is confident it will be the winner in such a "winner takes it all market", will tend to oppose compatibility.[40] Incompatibility is likely to result then, even if its competitors prefer compatibility. In many cases, such firms "will not be able to counteract and correct all incompatibilities introduced by an opponent, and, therefore, in such situations of conflict we expect that incompatibility wins" (Economides 1996, p. 687). Of course, there are goods

---

[40]   See Katz and Shapiro (1994), p. 111.

where firms can unilaterally impose some compatibility. However, compatibility is often not perfect. Especially in dynamic markets with frequent redesigns, competing firms will face a great deal of problems to keep up with compatibility. Moreover, a firm might possess patents or copyrights which might help to prevent other firms from establishing compatibility. (Certainly, if side-payments are feasible, a firm's desire as regards to compatibility might change, see Katz and Shapiro 1985, pp. 434 f.)

In a static setting, a similar argument can be found in Katz and Shapiro (1985) and Crémer, Rey and Tirole (2000). Assume there are two firms competing for "new unattached" buyers. The firms differ in the number of attached (perfectly committed) buyers, *i.e.* both firms possess an "*installed base*" of different sizes. Since network effects are involved if the firm's products are incompatible, the firm with the larger installed base (firm 1) offers better perceived relative quality than firm 2. Thus, firm 2 prefers compatibility. Firm 1, in contrast faces a tradeoff of two effects. On the one hand, since compatibility benefits all consumers, demand expands. On the other hand, there is a "quality differentiation effect": firm 1's perceived *relative* quality is augmented.[41] Depending on actual parameters, the increase in profits due to better relative quality may be larger than the losses through reduction of market demand. Hence, incompatibility may prevail even if compatibility is costless to achieve.

This argument also applies if compatibility is costly and firms can establish different levels of compatibility. If costs increase with the level of compatibility, the level of compatibility is lower than what would be socially optimal because the firm with the larger installed base does not take into account the impact of its compatibility decision on both its rival (whose profits increase with the level of compatibility) and consumers.[42]

To summarize, with imperfectly differentiated goods, two crucial effects determine the level of compatibility in network industries. First, with moderate network effects firms may have a (perhaps too) strong incentive to render their product compatible, because competition is lower in a compatibility regime. Second, with strong network effects where markets are likely to tip to one firm's variant, the likely winner of such a battle might prefer incompatibility because of the particular large rewards in a winner-takes-it-all market. Similarly, a quality differentiation effect might produce too little incentives for compatibility.

2.6   Empirical Evidence of Network Effects

Brynjolfsson and Kemmerer (1996) and Gandal (1994) investigate the market for spreadsheet programs. These studies make use of the fact that during the 1980's until early 1990's

---

[41]   See Crémer, Rey and Tirole (2000), p. 451.

[42]   See Crémer, Rey and Tirole (2000), p. 453.

Lotus 1-2-3 was by far the best seller. Using data from 1987 to 1991 and from 1986 to 1992, both studies indicate that consumers were willing to pay a significant premium for spreadsheets that are compatible with the Lotus platform (and, thus, provide for larger network benefits). Brynjolfsson and Kemmerer also show that there was a strong correlation between a program's total sales and its price: a one percent increase in a program's installed base corresponded to a 0.75 percent increase in its price.

Gandal, Kende and Rob (2000,1997) analyze positive feedback in the adoption of CD-players. Remember that, until introduction of the DVD-technology in 1996, the CD-technology was incompatible with any existing audio standard. Thus, according to the (indirect) network effect hypothesis we should observe positive feedbacks: sales of CD-players (hardware) depend on the variety of CDs (software) and vice versa. Estimating a dynamic model of technology adoption (based on date from 1985-1992), the authors corroborate this supposition. Dranove and Gandal (2001) examine the battle between DVD and DIVX.[43] They aim to find evidence for "vaporware" in this market. The phenomenon of vaporware has been identified by Farrell and Saloner (1986a). They argue that the timing of *announcements* of the introduction of a new incompatible technology may be crucial for its success in a standard battle. In order to slow down the adoption of competing, already introduced, incompatible technologies, firms may have incentives to announce the impending launch of their technology although the actual launch takes place later.[44] Such "wrongly" announced technologies are called vaporware. Dranove and Gandal find that preannouncements of DIVX did, in fact, slow down the adoption of the DVD technology.

Azoulay, Berndt and Pindyck (2000) examine consumption externalities in the US drug market. (Indirect) network effects in these markets may, among other sources, arise from the fact that a widespread use of a prescription drug may convey superior information to physicians and patients about its safety and efficacy. The authors focus on drugs employed for antiulcer/heartburn treatments and use data from the end of the 1970's to early 1990's. Their results indicate that past sales have contributed to the perceived value of a brand, but this effect is modest. However, its effect on the rate of diffusion turns out to be very significant.

In section 5, "discontinuous" adoption paths play a major role in our analysis. Such paths involve a jump of the adoption curve once a critical number of adopters has been reached.

---

[43] The DVD technology is similar to the conventional CD technology. One advantage of the DVD technology is that they provide for much larger data space. DVD-players can play CDs, however CD-players cannot play DVD discs (*i.e.* there is no upward compatibility). DIVX was supposed to become an, imperfectly compatible, competing standard to DVD, which was announced in 1997.

[44] For example, allegations of anticompetitive vaporware have been leveled against Microsoft and IBM. See, *e.g.*, Lemley and McGowan (1998a) for details.

There is much empirical support for this phenomenon. Besides some of the aforementioned studies, examples include Lange, McDade and Olivia (2001), for the Windows operation system, Economides and Himmelberg (1995), for the US fax market, Gabel (1993), for VCR, and Farrell and Saloner (1992), for US fax market and Color TVs.

To summarize, empirical studies do give significant evidence for network effects in many industries.[45] Unfortunately, these studies do not reveal whether we have too much or too little standardization and harmonization, neither on industry nor on international level. Besides, there are – to the author's knowledge – no empirical studies about network effects associated with social and legal norms. Of course, evidence for the presence of network effects within these areas may be seen by the fact that only one or a few sets of (network-) relevant social and legal norms tend to be adopted, especially within rather isolated nations or cultures.

---

[45] Other empirical studies include Koski (1999), Gowrisankaran and Stavins (1999), Goolsbee and Klenow (1999) and Saloner and Shepard (1995).

3. An Evolutionary Approach to Network Effects and Globalization

3.1 Introduction

Many standards vary across countries. On the one hand, this seems to indicate a desirable consequence of well functioning markets: if preferences vary across countries, one expects that goods are tailored to these different needs. On the other hand, however, with "globalization" different populations become more and more *integrated*. As the discussion in section 2.3 has shown, integration of incompatible networks – which is what globalization leads to if populations keep incompatible national standards – diminishes the network benefits of purely national standards. This leads us to the following questions: Do markets[46] react efficiently on increasing integration of countries ("globalization")? In particular, if goods are inherently incompatible, does globalization produce too much or too little harmonization? If globalization leads to harmonization, does it yield harmonization too early or too late? Does it imply that the "right" standard is implemented?

In order to answer these questions we introduce a simple framework that allows us to capture a wide class of network goods. We use an evolutionary model, which is based on pairwise interactions of agents from large populations, interpreted as countries. In each interaction, agents face a coordination game where both players' payoffs are always higher if both choose the "same" strategy. The choice of a strategy can be interpreted as an agent's adoption of a particular network good such as a specific social norm, a language, some type of software and so on.[47] The agents come from (homogeneous) populations that may differ in size and preferences of their members. An exogenous integration parameter is included, in order to analyze the impact of globalization.

The analysis demonstrates that, in competitive markets:

- Globalization does not necessarily implement harmonization, even if harmonization is efficient.

- *If* globalization implements harmonization, it does so too late from a social point of view. Thus, even modestly integrated markets are typically characterized by too much "variety" of standards.

- Globalization may produce harmonization with a socially inferior standard.

---

[46] Strictly speaking, there is no "classical" market for social norms and law provision. Since in our setup, individuals adopt such norms on an individual level, we keep the term as a proxy for the mechanisms behind.

[47] See also Holler and Wickström (1999), p. 100, and Kandori and Rob (1998), p. 33.

• The availability of double adoptions may support harmonization, even if they do not occur in (stable) equilibrium.

3.2 Review of Selected Literature

Law and Economics analyses incorporating network effects and heterogeneous preferences
of users (such as those dealing with institutional competition) often refer to the seminal paper
of Farrell and Saloner (1986b).[48] Consistent with our results, Farrell and Saloner find that
under these conditions, markets may or may not produce efficient outcomes. Outcomes may
be characterized by too much variety of networks. Since Farrell and Saloner do not assume
that variety is a "natural" starting point, the opposite may occur as well, *i.e.* harmonization
results even if variety is efficient.[49,50]

Our model departs from the analysis of Farrell and Saloner in several ways. Farrell and Saloner assume that agents possess different preferences for either variant's stand-alone value,
while network benefits are perceived as the same for both variants.[51] In contrast, in our setting, people's stand-alone benefits are the same for either good but network benefits differ.
This makes our analysis applicable to goods where benefits are mainly generated through interactions, *i.e.* where stand-alone values of goods play a minor role or are even negligible.
From an economic perspective, one could regard many legal or social norms and institutions
as such pure "network goods". That is to say, the value derived from "consuming" such
"goods" stems solely from the fact that under a given set of institutions, agents can interact
with each other more easily, which is an important common feature with classical networks
such as those of telephone and e-mail users.[52] For example, the value of clauses in a standard
contract is only generated when two parties are actually contracting.

Our modeling of consumer benefits also differs from de Palma, Leruth and Regibeau (1999)
and de Palma and Leruth (1996). In these models, consumers are differentiated according to
their valuation of the "network effect", yet this valuation is equal for either variant. The following model assumes quite the opposite. Agents from different populations value the "network effects" differently but the evaluation within each population is the same. Both assumptions are of course simplifications. De Palma and Leruth's version calls attention to the fact
that people do not benefit equally from enlargements of networks. For instance, the benefits
generated through the ability of speaking English are probably higher for a Spanish business-

---

[48] See, *e.g.*, Kahan and Klausner (1997) and Goerke and Holler (1995).

[49] Similar results yield Chou and Shy (1990, 1993) and Church and Gandal (1992, 1993).

[50] See section 2.2. for intuition on these results.

[51] A similar assumption is also made in Chou and Shy (1990, 1993) and Church and Gandal (1992, 1993).

man who travels intensively than for a baker who makes business only in his village. Our assumption, in contrast, emphasizes that Spaniards favor speaking Spanish in a conversation whereas people from the USA prefer English.

One contribution of our study is the explicit analysis of integration among countries. Also Belleflamme (1999) analyzes harmonization among different homogeneous communities. He assumes that communities face a Battle of the Sexes game. The effects of globalization are incorporated through the assumption that the payoffs in the coordinated outcomes (which represent harmonization) are larger the higher the integration among the communities is. Although, in the end, this applies to our setting as well, our approach is more direct. Instead of adjusting the "well-being" of the populations, the degree of integration only affects the likelihood of individuals being matched with members from each population.

Another characteristic of our evolutionary setup is that network effects are generated endogenously rather than to be exogenously assumed.[53] This is also true for the approaches of Chou and Shy (1993, 1992) and Church and Gandal (1993, 1992) as well as for Matutes and Regibeau (1992, 1988) and Economides (1989). In these models, network effects evolve through the exogenous assumption that (some) consumers benefit from variety of complementary goods.[54] Our approach, in contrast, is based on the assumption that interacting agents benefit from choosing the same strategy (compatible network good).

Other contributions dealing with network effects in an evolutionary setting are Andreozzi (2001) building upon Holler and Wickström (1999), Peters (1997), Kandori and Rob (1998) and Kandori, Mailath and Rob (1993), Ellison (1993) and An and Kiefer (1995). The explicit analysis of integration among populations distinguishes our model from these papers. Peters (1997) and Kandori, Mailath and Rob (1993) concentrate on one population. Kandori and Rob (1998) analyze "bandwagon properties" of a wide class of evolutionary games. They differentiate between the "global" and "local" interaction approaches. In the former, an agent is matched with any of the other agents with the same probability. In contrast, within the local interaction approach, as introduced by Ellison (1993) and followed by Ann and Kiefer (1995), agents mainly interact with their "neighbors". Ellison (1993) and Ann and Kiefer (1995) find that outcomes in such settings are much less dependent on historical events. This is due to the fact that even small mutations may induce some "domino-effect". Although our modeling

---

[52] See our discussion in section 2.1.

[53] See also Kandori and Rob (1998), p. 31.

[54] See section 2.1 for a brief discussion of the approaches of indirect network effects of Chou and Shy and Church and Gandal. In Section 6, the "mix and match" models of Matutes and Regibeau and Economides are discussed. These models assume that a share of consumers prefers "mixed" systems, thus these consumers benefit if networks are enlarged through compatibility.

seems to be located between the "global" and "local" approaches, historical events play a major role in our analysis.

Holler and Wickström's (1999) approach includes heterogeneous preferences and is therefore related to ours. While they assume that preferences are continuously distributed within one population, we focus on a discrete heterogeneous set of homogeneous populations, which allows to explicitly analyze the impact of integration among populations of different sizes. In other words, our analysis concentrates on the case where agents from different populations "prefer" different standards (given the same network size). Of course, this is not to say that preferences for network goods always differ across nations. Admittedly, the assumption that consumers' preferences differ among populations while being homogeneous within a population might appear unrealistic at first glance. Preferences, however, are often created by specific investments, experience or familiarity. And since we deal with goods, which are characterized by considerable network effects, harmonization has often already occurred on a *national* level. So, people from one population use only one kind of network good from which they receive experience and familiarity. Think of the strong preferences of the British for their own measurement or currency systems.

The remainder of this essay is organized as follows.[55] First, we give a brief introduction into the structure of the basic model (section 3.3). Afterwards, in section 3.4, equilibria are identified. In section 3.5, we compare the stable equilibria with socially desirable outcomes and ask whether increasing integration produces desirable outcomes. Section 3.7 tries to elucidate some major factors behind the results in order to justify some assumptions of the specific evolutionary setting. Section 3.8 extends the analysis and allows agents to adopt both "standards". In section 4, the second part of this essay, policy measures are analyzed that are deemed to produce harmonization.

3.3   The Model Setup

3.3.1 Populations

Our world consists of two homogeneous populations, *pop*1 and *pop*2. The populations are defined by the preferences of their members. That is, within each population the preferences of the members are all the same while the preferences of a member of *pop*1 are different from the preferences of a member of *pop*2. The sum of the sizes of both populations is normalized to 1. The fraction of individuals that belong to *pop*1 is exogenously given and denoted by $\sigma \geq 0.5$.

---

[55]   Sections 3.1-3.7 build up on Berndt and Simmering (2001).

3.3.2 Interaction Games

The members of the two populations constantly interact pairwise and choose one of the two available strategies, S_1 and S_2, for interaction. S_1 and S_2 may represent the adoption of a specific network good (or "standard") such as a language, some specific software, a specific contract term, a legal base of a contract, etc. If any two agents interact with each other *without using the same* standard, their payoffs are normalized to 0. An interaction of two players using the *same* standard always yields higher payoffs to both agents. An individual from *pop i* prefers interaction under his "domestic" standard, S_i, which gives her a payoff of $a_i>1$, compared to a payoff of 1 when the interaction takes place under the alien standard, S_j. Note that the agents' payoffs *do not depend on the affiliation* of the other player but *only* on the chosen strategies. See Figure 8 for the payoff matrices.

Figure 8
Payoff matrices for interactions

Payoff matrix *pop1*

Other player

	S_1	S_2
S_1	a_1	0
S_2	0	1

Player from *pop1*

$a_1>1$

Payoff matrix *pop2*

Other player

	S_1	S_2
S_1	1	0
S_2	0	a_2

Player from *pop2*

$a_2>1$

In sections 2.3 and 2.5, we have outlined that compatibility is an important parameter in networks. What does compatibility mean in our framework? The fact that "uncoordinated" interactions, *i.e.* (S_1,S_2) and (S_2,S_1), yield equal payoffs may represent *perfect incompatibility*. In contrast, if payoffs with (S_i,S_j) coincided with (S_i,S_i), then standards would be *perfectly compatible*. For example, if an agent from *pop*1 adopts S_1, he always receives a payoff of a_1, independently from the strategy chosen by his "opponent". Obviously, in such a state of perfect compatibility, network effects would not occur in the first place.[56] (Of course, harmonization is never beneficial in such a situation. It would only reduce the goods' variety without producing any benefits in return.)

We also noted in section 2.3 that standardization (making S_1 and S_2 more compatible) often requires some harmonization (making S_1 and S_2 more equal). In the matrix in Figure 8, this

---

[56] See also section 2.3.

would be equivalent to an increase in the payoffs received with the uncoordinated outcomes, *i.e.* (S_1,S_2) and (S_2,S_1), to the detriment of the payoffs with the coordinated outcomes, *i.e.* (S_1,S_1) and (S_2,S_2), respectively. For the time being, we keep the assumption that strategies are *inherently perfectly incompatible*, as given in the matrices in Figure 8. Only in an extension of the model in section 3.8, the effects of variations of compatibility are studied.

Another crucial assumption of the model is that the payoffs remain constant over time. Especially, it is assumed that there is no "sponsor" of either standard, for example firms who exclusively offer associated goods. Such a "sponsor" may have incentives to strategically vary prices or other parameters in order to make his standard prevail (see sections 5.2.2 for details).

3.3.3 Integration

Members of both populations may interact with both a member of their own or a member of the other population. The opponents for an interaction are taken randomly from both populations; the probabilities for cross-border interactions derive from an integration parameter, which is given exogenously. This assumption might not be innocent, as the integration level may also depend on the general level of harmonization of standards between countries. However, this effect should be negligible if we apply the analysis to a *single* standard only. Furthermore, recall that payoffs in the uncoordinated outcomes are *normalized* to zero. Obviously, many cross-border interactions, even under different standards, are still beneficial or even preferable. For example, business between firms from Germany and the USA does take place even though many national standards differ (*e.g.* language, measurement systems, paper format, etc.).

The level of integration among the populations is indicated by ω. $1-\omega$ stands for the probability that an individual interacts on the "domestic market", which only comprises agents of the same affiliation. With probability ω this agent interacts on the "world market" where he finds members of both populations, relative to the population sizes. For example, if $\omega=0$, an agent from *pop*1 always interacts with another agent from *pop*1 (one of his "compatriots"). For $\omega=1$, the probability that an agent from *pop*1 interacts with a compatriot depends only on the relative sizes of *pop*1 and *pop*2, *i.e.* the likelihood for each agent of being matched with someone from *pop*1 is σ and with someone from *pop*2 is $1-\sigma$.[57] Thus, one can say that $\omega=1$ resembles "perfect" integration.

---

[57] For simplification, ω is equal for both populations.

Figure 9

The world market and domestic markets

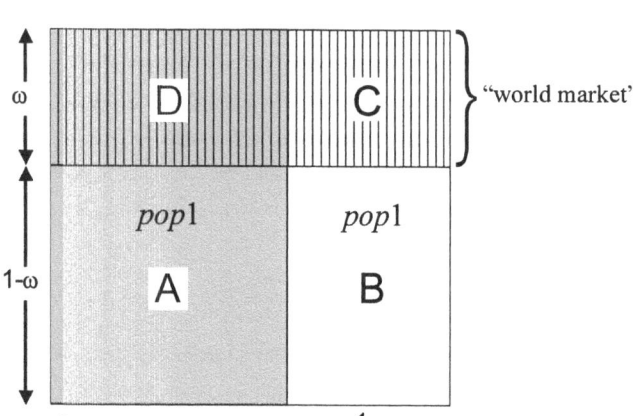

For illustration of the intermediate values of ω consider Figure 9. A member of *pop*1, for instance, can interact with a member of *pop*1 on the "domestic market" (area **A**), a member of *pop*1 on the "world market" (area **D**), and with a member of *pop*2 on the "world market" (area **C**). Thus, area **A+C+D** resembles the whole "market" for a member of *pop*1. Accordingly, the probability of a member of *pop*1 interacting with a compatriot is given by the fraction $\dfrac{\mathbf{A+D}}{\mathbf{A+C+D}}$. The probability of interacting with a member of *pop*2 (an "alien") is $\dfrac{\mathbf{C}}{\mathbf{A+C+D}}$. Other probability values derive analogously.

A state of the game is described by a strategy profile (φ_1,φ_2), where $\varphi_i \in [0,1]$ denote the proportion of individuals from *pop i* adopting S_1. The remaining fraction in that population ($1-\varphi_i$) uses S_2. The variables φ_i, σ, and ω determine the probability that an individual from *pop i* interacts with an individual who uses S_j, denoted by p_{ij}. Formally,

$$(E1)\ p_{ij}(\varphi_1,\varphi_2)=\begin{cases}(1-\omega)\varphi_i+\omega[\sigma\varphi_1+(1-\sigma)\varphi_2] & \text{for } j=1\\[2mm](1-\omega)(1-\varphi_i)+\omega[\sigma(1-\varphi_1)+(1-\sigma)(1-\varphi_2)] & \text{for } j=2\end{cases}\quad i,j=1,2$$

For illustration, consider p_{11}. With probability $1-\omega$, an individual from *pop*1 does not leave the "domestic market". The probability of interacting with a player using S_1 on the domestic market of *pop*1 depends only on the fraction of his compatriots using S_1. This explains the first term of p_{11}, $(1-\omega)\varphi_1$.

With probability ω, an individual from *pop*1 interacts on the "world market". This market consists of agents from both populations, relative to their sizes, σ and $1-\sigma$. Thus, for the agent of *pop*1, the probability of meeting a compatriot on the "world market" is σ. That agent uses S_1 with probability φ_1. With a probability of $(1-\sigma)$, the agent of *pop*1 interacts with an individual from *pop*2. Of these, φ_2 adopt S_1. Hence, the overall probability of interacting with a S_1-user on the "world market" makes up the second term of p_{11}, $\omega[\sigma\varphi_1+(1-\sigma)\varphi_2]$.

Note that the probability of being matched with someone adopting the same strategy on the *world market* does not depend on *who* adopts that strategy but only on the *total* number of agents. In contrast, the overall probability does depend on the distribution of strategies within the populations. Of course, if all agents on the domestic market adopt S_1, the probability of being matched with an agent adopting S_1 on the domestic market is 1.

3.3.4 Fitness

As noted before, we normalized the payoff of interactions under different standards to 0. Thus, the fitness f_{ij} (expected payoff) of a strategy j with respect to a member of population i is the product of the probability of interacting with an individual using the *same* standard times the payoff that results from such an interaction. If the probability of interacting with someone who uses the same standard is zero, then the fitness of the adopted strategy becomes zero. The fitness of a strategy increases (proportionally) with the probability of interacting with others who adopt the same strategy. Due to random matching, this probability, in turn, correlates with the *fraction* of agents who adopt the same strategy.[58] Thus, the strategies exhibit network effects.

As described in Figure 8, the significance of network effects depends on the preferences of the individuals. For a member of *pop i*, each interaction under the preferred standard S_i yields a payoff that is a_i times higher than the payoff under the alternative standard. Since we have normalized payoffs under different standards to zero, the fitness that a strategy yields to agents is simply the product of two factors: one factor is the probability of interacting with an agent who uses the same strategy. The other factor is the resulting payoff from interactions under that strategy. This payoff can either be a_i, if the preferred standard is the chosen strategy, or 1, if the alternative standard is used. This leads to the fitness functions f_{ij} given in (E2), where i denotes the individual's affiliation and j denotes the adopted strategy.

---

[58]   Subject to the level of integration, of course.

(E2) $$f_{ij}(\varphi_1,\varphi_2) = \begin{cases} p_{ij}(\varphi_1,\varphi_2)a_i & \text{for } i = j \\ p_{ij}(\varphi_1,\varphi_2) & \text{for } i \neq j \end{cases} \quad , \ i,j = 1,2$$

The fitness functions (E2) are illustrated for the case of $\omega=1$ (perfect integration) in Figure 10. In this specific case, the strategies' fitness only depends on the *total* fraction of S_1-users. Denote this fraction π:

(DI) $$\pi := \varphi_1\sigma + \varphi_2(1-\sigma)$$

We derive the following fitness functions:

(E2')
$$f_{11}(\pi) = \pi \cdot a_1$$
$$f_{12}(\pi) = (1-\pi)$$
$$f_{21}(\pi) = \pi$$
$$f_{22}(\pi) = (1-\pi) \cdot a_2$$

These are plotted in Figure 10. The fitness functions f_{11} and f_{12} pertain to agents from *pop*1, while f_{21} and f_{22} belong to members of *pop*2. As we can see in (E2'), a strategy's fitness depends on the fraction of agents using the same strategy. Since benefits are derived only from interactions and the available standards are perfectly incompatible, both $f_{11}(0)$ and $f_{21}(0)$ are 0. When all agents adopt S_1, *i.e.* $\pi=1$, f_{11} and f_{21} reach their maximum value. Due to different preferences, this maximum fitness of S_1 is greater for members of *pop*1 ($f_{11}(1) = 1\cdot3 = 3$) than for members of *pop*2 ($f_{21}(1) = 1$). The fraction of users of S_2 is given by $1-\pi$. Thus, $f_{11}(1)=f_{21}(1)=0$. At $\pi=0$, all individuals adopt S_2. Consequently, the highest fitness values of S_2 are observed, *i.e.* $f_{12}(0)=1, f_{22}(0)=4$.

Figure 10

Fitness functions with perfect integration (ω=0)

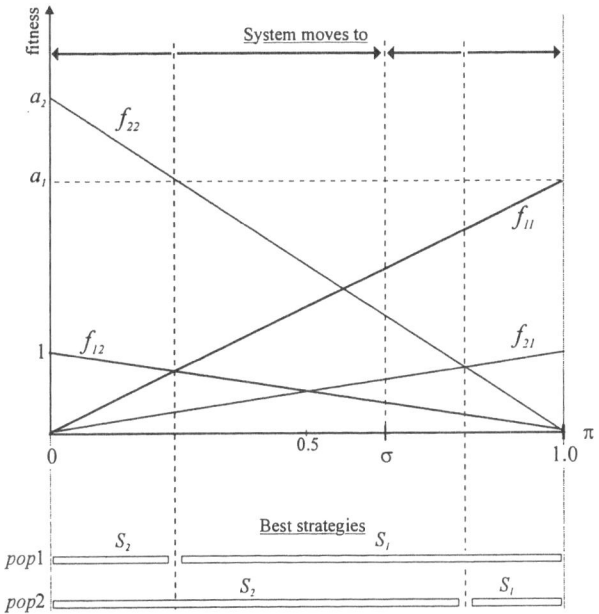

Exogenous parameters: $a_1 = 3$, $a_2 = 4$, $\sigma = 0.7$, $\omega = 1$.

Since f_{ij} are functions of φ_1 and φ_2, a graphical plot is, in fact, a three-dimensional plain. That is, there are in general two values for π which identify a unique strategy profile, (φ_1, φ_2). At $\pi=0$ the strategy profile is well-defined as $(\varphi_1, \varphi_2) = (0,0)$. Here we have harmonization with the "world standard" S_2, *i.e.*, all agents use S_2. At harmonization with S_1 ($\pi=1$) the strategy profile is well-defined as $(\varphi_1, \varphi_2) = (1,1)$. In order to interpret values for π between 0 and 1, we make the simplifying assumption that transition from $(\varphi_1, \varphi_2) = (0,0)$ to $(\varphi_1, \varphi_2) = (1,1)$ occurs in the following way. First, members of *pop1* start switching to their preferred standard S_1, thereby increasing φ_1 up to the point where $(\varphi_1, \varphi_2) = (1,0)$. Then, members of *pop2* start to change their strategy, which leads to an increase in φ_2 until the point $(\varphi_1, \varphi_2) = (1,1)$ is reached. This is only a simplification that helps to locate some points at the easy-to-read two-dimensional graph of Figure 10. It does not apply to the formal analysis of the model.

Under these simplifying assumptions, $\pi=\sigma$ corresponds to $(\varphi_1,\varphi_2) = (1,0)$. At this point, each agent is using his "preferred" standard, *i.e.* players from *pop*1 use S_1 and players from *pop*2 use S_2. We refer to this state as "variety". In contrast to states with harmonization, where all agents use either S_1 or S_2, the actual fitness values at variety do depend on the degree of integration. For example, the parameter values used in Figure 10 produce $f_{11}(0.7) = 2.1$ and $f_{22}(0.7) = 1.2$ with $\omega=1$, while $f_{11}(0.7) = 2.55$ and $f_{22}(0.7) = 2.6$ with $\omega=0.5$.

A brief comment on absolute size effects

Note that our modeling implies that *absolute* size effects do not play a role for network benefits. Formally, $f_{i1}(\varphi_1=1,\varphi_2=1,\omega=1)$ equals $f_{i1}(\varphi_1=1,\varphi_2=0,\omega=0)$. Analogously, this holds for S_2. The benefits with a particular strategy depend on the distribution of that strategy only within the set of agents that an agent actually interacts with. For instance, the fitness of S_1 for an agent from *pop*1 is the same for perfect integration where all individuals use S_1 and for perfect isolation where only individuals from *pop*1 use S_1. This resembles the reasoning in section 2.3, which has stated that network benefits decrease if the integration of populations increases, given that both populations stick to their incompatible standards and given that the individuals' intensity of use of network links remains constant. In other words, we assume that the *frequency* of social interactions that an individual enjoys is the same whether he lives in an isolated country or in an international environment and whether he lives in a small or in a large country.

It clearly depends on the specific environment whether this assumption is an appropriate representation. Consider, for example, two isolated countries that *only* differ in their size, say country 1 has one million and country 2 has five million residents. Is it reasonable to assume that a resident of country 2 derives more benefits from his language than someone from country 1? Probably not – such an assumption would imply that people living in a larger country conduct more conversations than people from a smaller country.

3.4 Existence of Equilibria

The existence of equilibria is based on the assumption that relatively fitter strategies attract more users during time. If a strategy yields higher fitness than another strategy to agents from pop i, the share of the former strategy increases relatively to the share of the latter in *pop i*.[59]

3.4.1 Isolated Populations ($\omega=0$)

Let us first consider a perfectly isolated population ($\omega=0$), say *pop*1. Since interactions only occur between members of the same population fitness of S_1 and S_2 are:

(E2'')
$$f_{11}(\varphi_1) = \varphi_1 a_1$$
$$f_{12}(\varphi_1) = 1 - \varphi_1$$

A graphical representation offers Figure 11.[60] Note that contrary to Figure 10 the abscissa represents φ_1 instead of π because agents only interact with compatriots, *i.e.* neither the populations' relative sizes nor the distribution of strategies within *pop*2 affect the fitness of the strategies.

As the plot reveals, there are two stable equilibria and one unstable equilibrium, applying the ESS-concept.[61] Harmonization with each strategy is a stable equilibrium. The interior one, which is determined by the intersection of f_{11} and f_{12}, is unstable. Any small invasion of mutants makes either one strategy fitter than the other, shifting the system away.[62]

Starting at a state left of the intersection of f_{11} and f_{12}, more and more individuals switch from S_1 to S_2, and so the population ends up with harmonization with S_2. In contrast, if the fraction of agents adopting S_1 is larger than at the intersection of the curves, the population will eventual enjoy harmonization with S_1.[63] Table 2 summarizes.

---

[59] This is a consequence of a Malthusian process, which is a standard assumption in evolutionary game theory, see, *e.g.*, Mailath (1998) and Holler and Illing (1996). See also section 3.7 below.

[60] Figure 11 corresponds to Figure 3 in Peters (1997), p. 350.

[61] This concept was introduced by Maynard Smith and Price (1973).

[62] Formally speaking, an invasion of mutants is a small (asymmetric) change in the strategy profile. So, this equilibrium is not stable, since the system will move away as soon as some very few agents change their strategy.

[63] See also Peters (1997), p. 350.

Figure 11

Fitness functions with isolated pop1

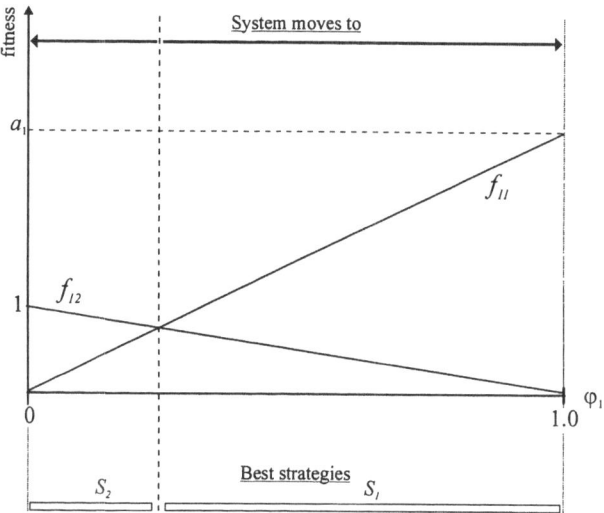

Exogenous parameters: $a_1 = 3$, $\omega = 0$.

Table 2

Stable equilibria in isolated population

	Harmonization with S_1	Harmonization with S_2
(φ_1)	**(1)**	**(0)**
f_{11}	a_1	0
f_{12}	0	1

☐ = fitness pertaining to the strategy that is played at the equilibrium

3.4.2 Integrated Populations ($1 \leq \omega < 0$)

For illustration of the case of perfectly integrated populations ($\omega = 1$), consider Figure 10 again. Starting from a state left of the intersection of f_{12} with $f_{11,}$ more and more agents of $pop1$ switch to S_2 because its fitness is larger for them. Since agents of $pop2$ stay with S_2, the populations end up with $\pi = 0$ where all members of both populations use S_2. For all states between the intersection of f_{12} with f_{11} and f_{21} and f_{22} the system moves to variety ($\pi = \sigma$). States right of the intersection of f_{21} with f_{22} will lead eventually to harmonization with S_1 ($\pi = 1$).

Both harmonization equilibria, $\pi = 0 \Leftrightarrow (\varphi_1, \varphi_2) = (0,0)$ and $\pi = 1 \Leftrightarrow (\varphi_1, \varphi_2) = (1,1)$, respectively, are stable, applying the ESS-concept. Two unstable internal equilibria also exist, which correspond to (φ_1, φ_2) at the intersections of f_{11} with f_{12} and f_{22} with f_{21}. These equilibria are not stable since any "invasion of mutants" turns one strategy fitter than the other one, and so more and more individuals are going to use that strategy, which will take the system either to the left or to the right. This also implies that only pure strategies are played in any stable equilibrium.[64]

A third stable equilibrium *may* exist at "variety" where (φ_1, φ_2) = (1,0), which corresponds to $\pi = \sigma$ given the simplifying assumptions for the graphical illustration. For the parameters chosen in Figure 10, variety is a stable equilibrium because the fitness of the domestic strategies is larger in either population.

Strategy profiles (φ_1, φ_2) and resulting fitness values (f_{ij}) of the potential three stable equilibria for the general cases are summarized by Table 3.

---

[64] Note that this is independent from the simplifying assumption made for Figure 10. If for any strategy profile both strategies yield equal fitness to agents from *pop1* and some agents from *pop2* use S_1, this state cannot be an equilibrium because S_2's fitness must be larger for agents of *pop2*. This implies that, in stable equilibria, agents only adopt pure strategies.

Table 3

	Harmonization with S_1	Variety	Harmonization with S_2
(φ_1, φ_2)	**(1,1)**	**(1,0)**	**(0,0)**
f_{11}	a_1	$[(1 - \omega) + \omega\sigma]\, a_1$	0
f_{12}	0	$\omega(1 - \sigma)$	1
f_{21}	1	$\omega\sigma$	0
f_{22}	0	$[(1 - \omega) + \omega(1 - \sigma)]\, a_2$	a_2

☐ = fitness pertaining to the strategy that is played at the equilibrium

3.4.3 Harmonization

As Table 3 indicates, harmonization with both standards, *i.e.* $(\varphi_1,\varphi_2) = (0,0)$ or $(\varphi_1,\varphi_2) = (1,1)$, are stable equilibria for any values of a_1, a_2, σ, and ω. Even for an individual from the population that does not favor the prevailing standard "as such", her strategy's fitness is 1 in equilibrium while the other pure strategy's fitness is 0, and any mixed strategies' fitness is smaller than 1. Thus, applying the ESS-concept, these equilibria are stable (because they are not upset by a small invasion of mutants).

3.4.4 Variety

The strategies played at the potential variety equilibrium, *i.e.* $(\varphi_1,\varphi_2) = (1,0)$, are ESS, as long as the exogenous parameters produce fitness values that satisfy $f_{11} > f_{12}$ *and* $f_{22} > f_{21}$. Plugging in the values from Table 3 and rearranging terms gives us conditions (C4a) and (C4b), which are to hold simultaneously.

(C4a)
$$\omega(1 - \sigma) < \frac{a_1}{a_1 + 1}$$

(C4b)
$$\omega\sigma < \frac{a_2}{a_2 + 1}$$

From these conditions, we can derive the following proposition.

Proposition 3.1: If populations are equally large ($\sigma = 0.5$), variety is always a stable equilibrium.

Proof: Since $a_i > 1$ the fractions on the rhs of (C4a) and (C4b) are always greater than 0.5. Since $\sigma = 0.5$ and $\omega \leq 1$, the lhs is cannot be greater than 0.5. Thus, (C4a) and (C4b) always hold. □

The intuition behind this proposition is straightforward. If populations are equally large, the probability to be matched with an alien is never greater than with a compatriot. Since agents favor interactions under their domestic standard, they keep adopting it even if the populations are perfectly integrated.

In general, since $\sigma \geq 0.5$ and $\omega \leq 1$, (C4a) always holds. Thus, whether variety is a stable equilibrium does not depend on a_1, the preferences pertaining to the larger *pop*1. Whether or not (C4b) holds depends on the parameter values. If *pop*2 is relatively large, integration is low and/or their members' preferences for their domestic standard are strong variety is more likely to be a stable equilibrium.

3.5   Social Welfare

From the fitness values given in Table 3, we can calculate values for utilitarian social welfare corresponding to the three potentially stable equilibria. In our framework, fitness can be interpreted as the average payoffs that a strategy yields to its adopter. Assuming that players are risk-neutral (such that payoffs are linear in fitness), we simply have to multiply the fitness of a strategy by the size of the populations that adopt it. Adding up gives us:

(E3) Harmonization with S_1: $W_{S1} = \sigma a_1 + (1 - \sigma)$

(E4) Variety: $W_V = \sigma[(1 - \omega) + \omega\sigma]a_1 + (1 - \sigma)[(1 - \omega) + \omega(1 - \sigma)]a_2$

(E5) Harmonization with S_2: $W_{S2} = \sigma + (1 - \sigma)a_2$

Note, for derivation of existence and ESS-stability of equilibria, it would be sufficient to assume that interactions under domestic standards yield higher payoffs than under alien standards and that payoffs of both interacting agents are always larger when both agents adopt the same standard. In contrast, to establish utilitarian welfare values we need our assumption that payoffs under each alien standard and under different standards are the same for agents of both populations.

3.5.1 Welfare With Harmonization

Comparing utilitarian welfare, we find that harmonization with S_1 is socially more desirable than with S_2 iff $W_{S1} > W_{S2}$ or,

(C5) $$\sigma a_1 + (1-\sigma) > \sigma + (1-\sigma) a_2 .$$

Defining $\Delta a_i := a_i - 1$ as a measure of how much more an individual of *pop i* prefers her own standard compared to the alien standard, one can rearrange equation (C5) to derive:

(C5') $$\sigma_1 \Delta a_1 > \sigma_2 \Delta a_2 \quad \text{with } \sigma_2 = 1 - \sigma_1 = 1 - \sigma$$

The *lhs* of (C5') represents the difference of *pop*1's well-being if harmonizing with S_1 instead of S_2. The *rhs* shows the same for *pop*2. If the gain in *pop*1 is higher than the loss in *pop*2 utilitarian welfare with harmonization with S_1 is higher.

Note that ω does not appear in condition (C5'). Thus, *which* of the standards is the better harmonization candidate does not depend on the integration level, ω. This is a consequence of our assumption that interaction payoffs do not depend on the affiliation of players but only on chosen strategies.

3.5.2 Efficiency of Variety and Harmonization

We now compare welfare of variety and harmonization.

Proposition 3.2: a) If populations are perfectly isolated ($\omega = 0$), variety is always socially optimal. b) If populations are perfectly integrated ($\omega = 1$), harmonization is always socially optimal.

Proof: For variety to be the efficient outcome, (C6a) and (C6b) must hold simultaneously:

(C6a) $$W_{S1} < W_V$$

(C6b) $$W_{S2} < W_V$$

By inserting (E3) and (E4), $W_{S1} < W_V$ implies:

(C6a')
$$\omega < \frac{a_2 - 1}{\sigma(a_1 + a_2)}$$

Analogously, by insertion of (E4) and (E5), $W_{S2} < W_V$ implies

(C6a')
$$\omega < \frac{a_1 - 1}{(1 - \sigma)(a_1 + a_2)}$$

a) $\omega = 0$: Variety is socially desirable if conditions (C6a') and (C6b') hold simultaneously. Since the expressions on the *rhs* of these conditions are greater than 0, variety is always optimal if $\omega = 0$.

b) $\omega = 1$: Plugging in $\omega = 1$, (C6a') and (C6b'), after rearranging, become:

(C6a'')
$$\sigma a_1 - (1 - \sigma)a_2 < -1$$

(C6b'')
$$1 < \sigma a_1 - (1 - \sigma)a_2$$

Since (C6a'') and (C6b'') contradict each other harmonization either with S_1 or S_2 yields higher social welfare. □

The fact that variety is efficient for low levels of integration is obvious. With $\omega = 0$, any interaction occurs under coordination and both players adopt their preferred strategy. Harmonization would not improve network benefits in either population, but only reduce benefits from goods' variety. When populations become more integrated, more and more uncoordinated interactions take place. The intuition for why $\omega = 1$ always implies that harmonization is efficient follows from the fact that coordinated interactions always yield higher payoffs than uncoordinated ones.

Let us consider the effects of harmonization in more detail. If $\omega = 1$, harmonization with S_1 and S_2 change populations' well-being in comparison to variety as indicated by (E6a) and (E6b), respectively.

(E6a)
$$W_{S1} - W_V = a_1\sigma(1-\sigma) - a_2(1-\sigma)^2 + 1 - \sigma$$

(E6b)
$$W_{S2} - W_V = a_2\sigma(1-\sigma) - a_1\sigma^2 + \sigma$$

The first term in (E6a) represents the gains of *pop*1 due to the larger network of their domestic standard. The second term stands for the losses of *pop*2, which occur because they ditch their preferred standard. The third term represents *pop*2's benefits through adoption of S_1 with maximum network. (E6b) is to read analogously.

First, imagine populations are symmetric, *i.e.* σ=0.5 and a_1=a_2. What *pop i* gains with its domestic standard equals the losses of *pop j* through sacrificing their domestic one. Since payoffs of agents of *pop j* with the alien standard are positive, harmonization is socially desirable. In the asymmetric case (σ>0.5 or $a_1 \neq a_2$) harmonization is even desirable if payoffs with the alien standard were zero. As it can be seen in (E6a) and (E6b), for *one* standard, additional network benefits with the domestic standard always exceed losses of the other population with their domestic standard. Thus, harmonization either with S_1 or with S_2 is desirable.

Note that even *both* populations might benefit from harmonization. For instance, if populations are symmetric and $1 < a_i < 2$, i = (1,2), both populations are better off with harmonization with either standard than with variety. In this example, full exploitation of network effects to the costs of adopting the more "preferred" standard is beneficial for either population.

3.6 Does Globalization Lead to Efficient Harmonization of Standards?

Let us define globalization as a continuous increase in the integration of the populations over time. *Proposition* 3.1 suggests that variety may be a stable equilibrium even if the populations are perfectly integrated (ω=1). *Proposition 3.2*, however, has shown that harmonization is always efficient for ω=1. Thus, considering variety as a natural starting point, populations may get stuck with variety even if through globalization, harmonization becomes socially superior to variety. Moreover, variety may persist even if harmonization yields higher well-being in *both* populations.

Nevertheless, if one population is (relatively) small enough and the preferences are not too strong, harmonization does occur once the integration level passes some threshold. For instance, if σ=0.2 and a_1=a_2=2 the critical integration level is ω=0.83. *Proposition 3.3*, however, shows that *if* globalization implements harmonization, it always does so too late from a social point of view.

Proposition 3.3: Suppose $(\varphi_1, \varphi_2) = (1,0)$, *i.e.* the system is situated at variety. Further suppose that integration increases constantly over time and the costs of transition to harmonization are negligible. Then, if harmonization occurs, it always occurs too late from a social point of view.

Proof: According to conditions (C4a) and (C4b), and *Proposition 3.2* variety is stable and efficient if $\omega=0$. Since (C4a) always holds, increasing integration can only make (C4b) fail. That is, starting at variety, increasing integration can only yield harmonization with S_1, the domestic standard of the larger population.

pop2's well-being with variety (W_V^{pop2}) and with harmonization with S_1 (W_{S1}^{pop2}) are:

(E7)
$$W_V^{pop2} = (1-\sigma)[(1-\omega)+\omega(1-\sigma)]a_2$$

(E8)
$$W_{S1}^{pop2} = 1-\sigma$$

W_{S1}^{pop2} is independent from ω. W_V^{pop2} decreases strongly monotonically with ω:

(E9)
$$\frac{\delta W_V^{pop2}}{\delta\omega} = (\sigma^2 - \sigma)a_2 < 0$$

The *lhs* of (C4b) increases strongly monotonically with ω while the *rhs* is independent from ω.

Thus, we have to show, if $W_{S1}^{pop2} = W_V^{pop2}$, then (C4b) holds.

Define with ω^{CR} the critical value of ω such that $W_{S1}^{pop2} = W_V^{pop2}$ holds:

(DII)
$$\omega^{CR} := \omega \mid W_V^{pop2} = W_{S1}^{pop2}$$

Plugging in (E7) and (E8) yields

(E10)
$$\omega^{CR} = \frac{a_2 - 1}{\sigma a_2}$$

After inserting (E10) into (C4b) we obtain

(C7)
$$\frac{a_2 - 1}{a_2} < \frac{a_2}{a_2 + 1}$$

which always holds. Since *pop*1 is better off too if *pop*2 switches to S_1, the proposition is proven.   □

Two issues are responsible for the excessive stickiness of variety. The first one is a "collective action problem". Agents do not unilaterally switch as long as the fitness of their domestic standard is larger. Thus, they still stay with their domestic standard even if their benefits with the alien standard would be higher if *all* of them switched. The second reason for this is a (positive) externality that accrues to the alien population through the enlargement of their network. Hence, even if a population could coordinate a collective transition to the other standard, it would do so too late because their members do not take into account the increase in well-being of the alien population.

Since (C4a) always holds, increasing integration never yields harmonization with S_2, the standard of the smaller population. Social welfare might however indeed be higher when the populations harmonize with S_2. This poses the question whether it is possible that globalization produces harmonization with the "wrong" standard.

Proposition 3.4: Suppose $(\varphi_1, \varphi_2) = (1,0)$, *i.e.* the system is situated at variety. Further suppose that integration increases constantly over time. Then, if harmonization occurs, it might occur with the standard S_i although harmonization with S_j $(i \neq j)$ would produce higher social welfare.

Proof: To prove the proposition it suffices to give an example. Suppose $a_1 = 1.1$, $a_2 = 3$, $\omega = 1$, $\sigma = 0.8$. These values produce:

$W_{S1} = 1.08$
$W_{S2} = 1.4$
$f_{21} = 0.8$
$f_{22} = 0.6$

This implies that $W_{S2} > W_{S1}$ and (C4b) fails. Thus, harmonization occurs with the S_1 although harmonization with S_2 would yields higher welfare. (Since $\omega = 1$, $W_{S2} > W_V$. We do not need to calculate some critical ω because agents from *pop*1 never switch to S_2.)   □

Remember, situated at variety, agents of the larger *pop*1 never switch to S_2 because the probability of being matched with an alien is never greater than the probability of being matched with a compatriot. Since they prefer interactions under S_1, they never switch to S_2. This result also implies that if "inefficient" harmonization occurs, then it is to the detriment of the smaller population. For agents from the smaller *pop*2, at variety, f_{21} may very well be larger than f_{22}, even though $W_{S2} > W_{S1}$. A comparison of the welfare values of alternative harmonization states also accounts for the potential benefits of a *collective* switch of *pop*1 to S_2, which is conferred on both agents from *pop*1 and from *pop*2. In the above case, the externality on *pop*2 of a switch of *pop*1 to S_2 is larger than the externality *pop*1 can bestow on *pop*2.

3.7 Some Interpretations of the Assumptions

3.7.1 Darwinian Selection

The system's dynamic is generated by the natural appearing assumption that agents tend to switch to fitter strategies over time. Such adjustments resemble Darwinian selection: survival of the fittest. Evolutionary game theory has its origin in Biology. For example, animals that adopt successful strategies produce more offspring than animals that adopt less successful strategies. The offspring tend to adopt the strategy of their parents – *e.g.*, due to genetic inheritance – and so, the fractions of fitter strategies (or animals that adopt these strategies) grow.

In a social context, one assumes that agents somehow learn which strategies are successful. Evolutionary models in economics usually do not explicitly model the learning process that is involved. Instead, one assumes that individuals somehow tend to figure out, through experience and imitation, which strategies are currently more successful (fitter) than others (see Mailath 1999, p. 84). Although agents cannot directly observe fitness (as it is an expectation value) this assumption seems reasonable as long as the number of agents' observations of the payoffs that they themselves and other individuals receive is sufficiently large. In the end, it is only assumed that more individuals switch from less fit to more fit strategies than vice versa.

Since agents tend to adopt strategies that are best responses to the *present* distribution of strategies (strategy profile), agents are assumed to be myopic. In general, such myopic behavior appears particularly reasonable when there are lots of players and if the rules of the game are not obvious[65], which should apply to our game. To play the best response to a current state does not necessarily imply that "rational" behavior (perfect foresight) is different, though. In

---

[65] See Mailath (1999), p. 84.

fact, if it is costless to switch between strategies and agents always know the present state, then myopic behavior coincides with rational behavior in our game.

Note that imitation is somewhat critical in our model. Recall that the strategies' fitness depends on the affiliation of the agents. We have assumed that the share of strategies that are relatively fitter for agents of *pop i* grows in *pop i*. The same strategy may very well be the relatively less fit one and lose share in *pop j*. Thus, we have assumed that agents imitate selectively, *i.e.* they only imitate agents from their own population. This is an unproblematic assumption if integration level is low, as agents are matched with agents from their own population for the most times. With high integration, our assumption implies either that agents can differentiate between agents when they observe others' payoffs or that agents learn mainly through experience rather than through imitation.[66]

3.7.2 Conditional Strategies

In the previous sections, we assumed that the strategy set of the agents entails the choice of one out of two standards. If the agents know the affiliation of their opponents in advance, *i.e.* before they choose their strategy for interaction, the strategy set, however, includes conditional strategies $S^C = (S_{i|1}, S_{j|2})$, $i, j = (1,2)$, where $S_{i|k}$ signifies that an individual plays standard i when interacting with an individual from *pop k*. In fact, our assumption that agents imitate selectively may suggest the availability of such strategies.

As discussed in subsection 2.4, typically, network goods are durable. Thus, the adoption of a strategy or standard may require an investment.[67] For instance, people have to learn English before communicating in English, computer users have to buy and install a particular program before exchanging data with it, people have to buy a telephone and extension before making phone calls, etc. Payoffs in the interaction games are to be interpreted as being net of the associated costs such as for interest and depreciation.[68] However, due to sunk costs involved, such investments typically generate switching costs.[69] While this might alter the speed of the adjustment process it does not change the system's dynamic qualitatively.[70] It does, however, reduce the fitness of conditional strategies.

---

[66] See Berndt and Simmering (2001) for implications if agents imitate non-selectively.

[67] See also section 2.4.

[68] This remains a little vague here, as we do not explicitly consider the time in the model.

[69] See section 2.4.

[70] As discussed in the previous section, agents take a present state as given, *i.e.* they assume that the strategy profile observed in the "last period" remains unchanged today. Thus, if switching costs are not too large, agents still tend to switch to (presently) relatively fitter strategies. Such behavior is rational too if the adjustment process is slow relative to the discount factor (see also Kandori and Rob 1998, p. 35).

This is so because without harmonization, the adoption of a conditional strategy would require frequent switches of the chosen standard because an agent's opponents are chosen randomly. For example, let the system be situated at a state where all agents from *pop*1 adopt $S^C = (S_{1|1}, S_{2|2})$ and all agents from *pop*2 adopt $S_2 = (S_{2|1}, S_{2|2})$. This describes a state where agents of *pop*1 adopt their domestic standard when interacting with a compatriot and the alien standard when being matched with an alien. Agents of *pop*2 always adopt their domestic standard. Of course, this state can be a stable equilibrium. However, the presence of switching costs reduces the fitness of S^C by $2\sigma(1-\sigma)k$, if $\omega=1$, where k denotes the switching costs that are associated with a change of the adopted standard, and $2\sigma(1-\sigma)$ is the probability of such change.[71] If k is not too small and σ not too large, the fitness of conditional strategies such as S^C are always lower than fitness of unconditional ones. Thus, in those cases, agents never adopt such strategies, and so the original model with unconditional strategies applies, even if the agents know the affiliation of their opponents before they choose their standard for interaction.[72]

For illustration, consider a "battle of the forms": suppose, firms originating from two countries, say Belgium and England, do business regularly with each other. One important contract provision is the legal base of their contracts. Firms can either choose "English law" or "Belgian law". Assume that the resulting payoffs from the choice of the legal bases fit our payoff structure (as given in Figure 8). That is, the payoffs of the firms are always larger when both interacting firms have chosen the *same* legal base. However, Belgian firms prefer contracts based on Belgian law, while English firms prefer contracts under English law.

Adopting an "unconditional" strategy, a firm chooses either "English law" or "Belgian law" for interactions in general, irrespective of the affiliation of the specific firm they do business with. In contrast, when adopting a conditional strategy, a firm chooses English law when interacting with an English firm, and Belgian law when interacting with a Belgian firm, for example. However, suppose the adoption of a law requires an investment. For example, in order to adopt a specific law, firms must have sufficient expertise on the specific law and/or they must employ specific lawyers that are experienced with the law, etc. It is obvious that such investments involve considerable sunk costs. Thus, a conditional strategy is more costly than an unconditional one because it is costly to frequently switch between the adopted laws. If such frequent switches are too costly, then even if firms know the affiliation of the contract

---

[71]   It is the sum of the probabilities of a change from S_1 to S_2 and from S_2 to S_1. The probability of having adopted S_1 at last interaction is σ, the probability of a switch to S_2 is $1-\sigma$.

[72]   Even though one may believe that there are relevant cases where such switching costs are low enough such that conditional strategies are actually adopted, we omit a detailed analysis here.

partner in advance, they would adopt "unconditional" strategies, *i.e.* either English or Belgian law.

One might object that some firms could find it beneficial to adopt both laws *at the same time*, *i.e.*, firms maintain at the same time expertise in both laws and employ both lawyers that are familiar with English law and lawyers that are familiar with Belgian law. This is likely to allow for a more flexible choice of the legal base of contracts. On the other hand, however, it may be very costly to maintain expertise in both areas. Such strategies are the focus of the next section.

3.8 Extension: Double Adoptions – Production of Compatibility by Users

In the previous sections, we assumed that agents could adopt only one standard *at the same time*. As discussed in sections 2.3, if standards are not mutually exclusive it might however pay for agents to adopt both standards at the same time. The benefits of such double adoptions are an increase in compatibility to the standard adopted by one's opponents. Such phenomena might appear pretty weird. After all, agents acquire two (or more) variants of a good although these variants, at first glimpse, are substitutes. Nevertheless, such behavior is often observed. For instance, many people speak more than one language, many computer users have installed both MS Internet Explorer and Netscape Communicator, and many people carry several credit cards. At the beginning of the century, some "rich" customers even dealt with the coexistence of several incompatible US phone companies by getting linked to several of them.[73] These observations pose the question whether such phenomena occur temporarily during some adjustment process or can, in fact, be observed in a stable state. Moreover, how does the availability of such double adoptions affect the existence and desirability of equilibria?

3.8.1 Review of Selected Literature

De Palma, Leruth and Regibeau (1999) analyze the effects of double adoptions in a duopoly setting. Firms can design their products compatible or incompatible with each other. However, even in cases of incompatibility, consumers can always reap the benefits of compatibility by making double purchases, *i.e.* through joining both networks. Crémer, Rey and Tirole (2000) use a similar framework. They analyze "multihoming" of Internet Service Providers (ISPs) to Internet Backbone Providers (IBPs). Network effects in their setting arise through the ISPs' wish to have a high quality (perfectly compatible) link to as many other ISPs as pos-

---

[73] See de Palma, Leruth and Regibeau (1999), p. 211 (who refer to Frank Koppelmann).

sible. Through multihoming, ISPs achieve perfect compatibility of links to all ISPs that are linked to one of those IBPs they are linked to.[74]

Both models exhibit that double adoptions can occur in equilibrium. Unfortunately, none of them provides for a comparison of social welfare in the presence and the absence of the possibility of double adoptions. Crémer, Rey and Tirole stress that the possibility of users' double adoptions may increase the incentives of a dominant firm to reduce compatibility. Since users can provide for compatibility themselves, the negative effect on demand is lower if the firms design their goods less compatibly. De Palma, Leruth and Regibeau, in contrast, show that the equilibrium degree of compatibility tends to be higher with the possibility of double adoptions, even in cases where they do not actually occur in equilibrium. The reason is that the possibility of double adoptions promotes a symmetric equilibrium in which both firms favor a higher degree of compatibility.[75]

Another related contribution is Farrell and Saloner (1992), who investigate how the availability of converters affects performance of network markets. They assume that standards are mutually exclusive but the adoption of a converter increases the compatibility of the alternative standards. Thus, to buy a converter is very similar to adopting both standards. Farrell and Saloner (1992) show that if a converter is costly and provides only for imperfect compatibility, they tend to be used excessively. The reason for this is that the net externalities conferred on other users are greater when a user adopts S_i than if he adopts S_j and buys a converter. In an equilibrium where people use converters, more people use S_i than S_j.[76] Consequently, since compatibility with converters is imperfect, the externality bestowed through adoption of S_i is greater than through adoption of S_j plus a converter.[77] For the same reason, Farrell and Saloner further show that the availability of converters may preclude harmonization although harmonization is the efficient state *and* would be achieved when converters are *not* available.

An interesting coordination problem arises in Farrell and Saloner. Imagine, converters are not available and agents are situated in a state of variety, *i.e.* agents are divided into two

---

[74] ISPs are not directly linked with each other. Instead, ISPs get linked to an IBP. IBPs, in turn, are linked with each other. If these links are imperfectly compatible, then ISPs favor to have links to IBPs with a large base of (other) ISPs. This, in turn, may produce incentives for ISPs to get connected to several IBPs, *i.e.* for "multihoming".

[75] Since the incentives for compatibility can even become excessive, welfare effects can be positive or negative.

[76] This is due to their assumptions on consumer heterogeneity. Network benefit functions are equal for all consumers. Intrinsic preferences for alternative variants, in contrast, are assumed to be heterogeneous (equally distributed). Agents located at the boundaries have the strongest preferences for S_i and S_j, respectively. Thus, the agent located in the "middle" prefers S_i to S_j together with a converter, which implies that in equilibrium where converters are used, more agents use S_i than S_j together with a converter.

[77] See Farrell and Saloner (1992), p. 24.

groups where group i adopts S_i and group j adopts S_j. Now, converters become available. Agents of *pop* i do not buy converters if all agents of *pop* j do so, and vice versa, because the compatibility of the standards is the same whether group i, group j or both groups adopt converters. To give a double-adoption example, if all Germans speak *both* German and English, there is no incentive for Americans to learn German. Consequently, in equilibrium only one group learns both languages. But which group will actually do so? Farrell and Saloner "simply assume that the coordination problem *is* somehow solved"[78]. Eventually, one group *will* buy converters.[79]

Integrating the possibility of double adoptions into our evolutionary framework, we first offer a solution to the problem of which group of buyers actually chooses to make double adoptions (adopt converters). Furthermore we show, in contrast to Farrell and Saloner, that the possibility of double adoptions may actually support efficient harmonization – rather than precluding it.

3.8.2 The Strategy S_b

To incorporate double adoptions into our evolutionary framework, we must extend our payoff matrices and include the strategy S_b, which stands for double-adoption. An agent who adopts S_b enjoys (some) compatibility with both standards. We further assume that such an agent can differentiate between the affiliations of his opponents and chooses the interaction standard accordingly.[80] That is to say, if two interacting agents from *pop* i adopt S_b, both of them receive a payoff of a_i. If opponents originate in different populations and both adopt S_b we assume they receive, in average, $(a_i+1)/2$. Sticking to the language example, this convention accounts for the fact that a conversation, in most cases, is conducted in *one* language. Suppose a German talks with someone from England and both know either language. Our payoffs imply that within their conversation, either both speak German or both speak English. A conversation where the German speaks German and the English speaks English is hardly possible and thus excluded here. It is assumed that in such cases, the conversation's language is chosen randomly, which produces $(a_i+1)/2$.

Furthermore, we can expect that someone who knows German *and* English speaks English with someone who *only* learned English. Consequently, if an agent who adopts S_b is matched with an agent who adopts S_j, interaction takes place under standard j. Thus, similar to Farrell

---

[78]   Farrell and Saloner (1992), p. 19.

[79]   A similar phenomenon has already occurred in section 2.5, where we have investigated the compatibility incentives of firms in perfectly differentiated markets.

[80]   Switching between *adopted* standards is assumed to be costless.

and Saloner, our assumptions imply that agents *cannot* achieve *perfect* compatibility between the two standards.

The adoption of S_b is more costly than the adoption of a strategy facilitating the use of only one standard. We account for this natural assumption and deduct $c>0$ from the payoff received in each interaction an agents conducts with S_b. Our assumptions produce payoff matrices as displayed in Figure 12.

Figure 12

Payoff matrices for interactions including double adoptions

Other player (from *pop*1)

Player from *pop*1

	S_1	S_2	S_b
S_1	a_1	0	a_1
S_2	0	1	1
S_b	a_1-c	1-c	$a_1 - c$

Other player (from *pop*2)

Player from *pop*1

	S_1	S_2	S_b
S_1	a_1	0	a_1
S_2	0	1	1
S_b	$a_1 - c$	$1 - c$	$\frac{1}{2}(a_1+1) - c$

Other player (from *pop*2)

Player from *pop*2

	S_1	S_2	S_b
S_1	1	0	1
S_2	0	a_2	a_2
S_b	$1 - c$	$a_2 - c$	$a_2 - c$

Other player (from *pop*1)

Player from *pop*2

	S_1	S_2	S_b
S_1	1	0	1
S_2	0	a_2	a_2
S_b	$1 - c$	$a_2 - c$	$\frac{1}{2}(a_2+1) - c$

3.8.3 The Effects of Double-Adoptions

Instead of completely solving the model we want to demonstrate the following two results:

a) Consider variety with perfect isolation as the starting point and let there be globalization. Then we can predict *which* population, if any, adopts both standards (S_b). However, from a social point of view it is not always the "right" population that switches to S_b.

b) The possibility of double adoptions actually supports harmonization rather than precluding it – even though double adoptions do not occur in equilibrium.

Result a)

A natural starting point is perfect isolation and variety. Then, we let the integration level ω increase and ask which population adopts S_b first (if any). If the adjustment process is fast relative to the increase in integration, no agent of *pop i* has an incentive to switch from S_i to S_b if agents from *pop j* have started to do so. f_{ii} increases with each agent of *pop j* adopting S_b. Thus, consistent with Farrell and Saloner, in (stable) equilibrium, at most one population adopts S_b. Second, the fitness of S_b for agents of *pop j* is independent from the share of agents of *pop j* that adopts S_b or S_j. Hence, in (pure strategy) equilibrium, either all agents of *pop j* adopts S_b or none. This also parallels Farrell and Saloner.

Building on (E1) and (E2), *situated at variety or in a state where some agents of pop2 adopt* S_b, the fitness values of S_2, S_1 and S_b for agents of *pop2* are:

(E11a) $$f_{22}(\varphi_1 = 1, \varphi_2, \vartheta_2, \omega) = (p_{22} + p_{2b})a_2$$

(E11b) $$f_{21}(\varphi_1 = 1, \varphi_2, \vartheta_2, \omega) = p_{21} + p_{2b}$$

(E11c) $$f_{2b}(\varphi_1 = 1, \varphi_2, \vartheta_2, \omega) = (p_{22} + p_{2b})a_2 + p_{21} - c$$

p_{2b} denotes the probability that an agent of *pop2* is matched with someone adopting S_b.[81] ϑ_2 is the share of S_b within *pop2*.[82] As (E11a) and (E11c) confirm, if some agents of *pop2* switch from S_2 to S_b, the fitness of S_2 and S_b within *pop2* is unaffected. Consequently, in pure strategy equilibrium, either every or no agent of *pop2* adopts S_b.

---

[81]   Remember that p_{ij} denotes the probability of an agent from *pop i* to be matched with an agent adopting S_j (no matter from which population).

[82]   Recall that φ_i denotes the share of agents in *pop i* that adopt S_1.

Agents of *pop2* start to switch from S_2 to S_b if ω produces $f_{2b} - f_{22} \geq 0$. Hence, S_b will start to become adopted by agents of *pop2* as soon as $p_{21} \geq c$. At variety, the likelihood that an agent of *pop2* is matched with someone adopting S_1 is:

(E12) $p_{21} = \omega\sigma$

We can calculate analogous expressions for agents of *pop1*. Denote ω^{b1} and ω^{b2} the critical integration level where agents of *pop1* and *pop2* start to switch from S_1 to S_b and from S_2 to S_b, respectively. Following from (E11a,b,c), (E12) and analogous calculations for *pop1*, these are:

(E13a) $$\omega^{b1} = \frac{c}{1-\sigma}$$

(E13b) $$\omega^{b2} = \frac{c}{\sigma}$$

Since, by convention, $\sigma \geq 0.5$, $\omega^{b2} < \omega^{b1}$. Thus, *if* S_b is adopted, then agents from the smaller population will switch first.[83] If, as it seems reasonable, the adjustment speed is fast relative to the increase of the integration level, agents of the larger population never switch to S_b.

Of course, this is not always efficient. A benevolent dictator who decides which population switches to S_b would also account for the preferences within the populations – not only for the populations' relative sizes. Hence, we can set up the following proposition:

Proposition 3.5: Starting from ω=0, if globalization causes S_b to be adopted, then S_b will always be adopted by the smaller population. This may or may not be efficient.

Result b)

Now we want to show that the availability of double adoptions may facilitate efficient harmonization. To do so, we choose an example:

$$\sigma = 1 - \sigma + \varepsilon, \text{ where } \varepsilon > 0 \text{ is small}$$
$$a_1 = a_2 = 1.5$$
$$c = 0.5 - \varepsilon$$

---

[83] S_b will become adopted if $c \leq \sigma$. Otherwise, agents of *pop2* rather switch to S_1 than S_b.

These parameter values produce values of social welfare for $\omega = 1$:

$$W_{S1} \approx W_{S2} \approx 1{,}25; \ (W_{S1} > W_{S2})$$
$$W_V \approx 0.75$$
$$W_{Sb} \approx 1{,}125$$

W_{Sb} represents a state where each agent of *pop2* adopts S_b and each agent of *pop1* adopts S_1. Since $c = 0.5$-ε and *pop1* is just larger than *pop2*, harmonization with S_1 is the efficient state. (Of course, if c=0 then W_{Sb} would be efficient.) As populations are (almost) equally large, globalization would not implement harmonization if S_b were not available (see *Proposition 3.1*).

Figure 13

Implementation of harmonization through double adoptions

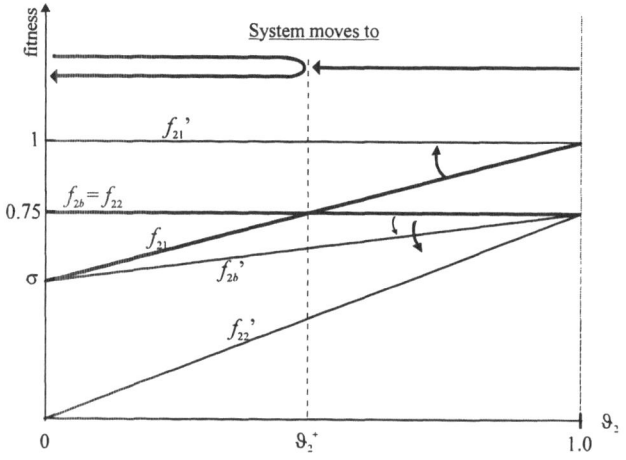

Exogenous parameters: $a_2 = 1.5$, $\omega = (0.5$-$\varepsilon)/(0.5$+$\varepsilon)$, $\sigma = 1$-σ+ε, $c = 0.5$-ε.

Again, we let the integration increase, starting from ω=0. Since c>0, agents from both populations use their domestic standard, S_1 and S_2, respectively (*i.e.* ϑ_2=φ_2=0). With increasing ω, S_b becomes more and more fit relative to S_1 and S_2, respectively. In Figure 13, the abscissa measures ϑ_2, the proportion of *pop2* adopting S_b; the vertical axis gives the fitness. Based on φ_1=1, *i.e.* each agent originating in *pop1* adopts S_1, the figure depicts various fitness functions for $pop2$.[84] Figure 13 assumes that $\omega = \omega^{b2} = (0.5\text{-}\varepsilon)/(0.5\text{+}\varepsilon)$, *i.e.* the critical integration level has just been reached where double adoption yields as much benefits as adoption of the domestic standard to agents from $pop2$. Let ω marginally increase, which turns S_b slightly fitter than S_2. This launches an adjustment process. More and more agents of $pop2$ switch from S_2 to S_b, shifting the system to the right. f_{22} and f_{2b} remain unchanged. However, f_{21} increases, because the probability of being matched with someone who is compatible with S_1 increases. When ϑ_2 reaches ϑ_2^+, the critical number is attained where the fitness of S_1 and S_b are equally large.[85]

Since ω is slightly larger than ω^{b2}, at ϑ_2^+, $f_{22} < f_{2b} = f_{21}$. This triggers more agents who still adopt S_2 to switch to S_b and to S_1. This has two effects. First, due to the increase in ϑ_2, the system moves to the right. Second, through the increase in the share of S_1-adopters, the curves rotate. While f_{21} rotates clockwise, f_{2b} and f_{22} rotate counterclockwise. (The figure only depicts the fitness function for the two extreme cases, f_{2i} belong to $\varphi_2=0$ and f_{2i}' belong to $\vartheta_2+\varphi_2=1$). Both effects turn f_{21} larger than f_{2b}.[86] Hence, S_1 attracts more and more agents. Eventually, all remaining S_2- and S_b-adopters will switch to S_1. And so, S_b will die out again. Thus, we can set up the following proposition:

Proposition 3.6: The possibility of double-adoptions may implement efficient harmonization, even though double-adoptions do not occur in equilibrium.

3.8.4 Illustration

To give a story, imagine that more and more people in Germany learn English as a second language. Since this increases the total number of English speaking people, the incentive to learn German decreases for Americans. However, not only the Americans' motivation to do so decreases. At some point, even people from Germany – even if they intrinsically favor speaking German – might not want to learn or keep German. If preferences of Germans to speak German in a conversation are not too strong, the additional network benefits of English – produced by Germans who learned English as a *second* language – causes people from

---

[84]   In view of the fact that ω^{b2} is almost 1 we can approximate the fitness functions with linear curves.

[85]   If $\omega=\omega^{2b}$, this would be an equilibrium – an unstable one, though.

[86]   To be precise: this, in turn, triggers both S_b- and S_2-adopters to switch to S_1. Those agents who switch from S_2 to S_1 further increase f_{21} through further rotating the curve clockwise. Those who switch from S_b to S_1 trigger two effects. On the one hand, f_{21} increases since the curve rotates clockwise. On the other hand, f_{21} decrease because ϑ_2 shrinks. In any case, since f_{21} rotates clockwise, the intersection of f_{21} and f_{2b} shifts to the left. Repeating this process makes S_b eventually die out, and the population ends up with all of them adopting S_1.

Germany to learn English rather than German or German *and* English. It does not pay to bear the costs of learning two languages. This, in turn, augments the network benefits of the English language even more, while network benefits of German decrease. In the end, this leads to a state where everyone speaks English and no one German (and no one both languages).

Our result is diametrically opposed to Farrell and Saloner (1992) who stress that the availability of converters tends to preclude efficient harmonization. What is the reason? One reason is that variety is stickier in our framework than in the model of Farrell and Saloner. In our model, harmonization and variety typically co-exist as equilibrium states if double adoptions are not possible. In their model, without converters, harmonization and variety are unique equilibria, *i.e.* depending on the strength of network effects, either harmonization or variety occurs (this is due to exogenous assumption of network effects and the modeling of the consumer preferences). In cases where harmonization does *not* exist and users adopting S_2 buy converters, this enhances network benefits of S_1, too. However, if converters increase network benefits of S_1 by so much that each user would adopt S_1 rather than S_2 together with a converter, then converters are not adopted in the first place. Thus, variety occurs – not harmonization. In the other case where harmonization exists, variety does not. Consequently, the availability of converters can only preclude harmonization – not implement it.

The latter phenomenon cannot occur in our framework. If c is sufficiently low and preferences for the domestic standard are strong enough, S_b completely and permanently prevails in *pop*2. Thus, to give an answer to our question in the introduction, double adoption may be both – a temporary or a permanent phenomenon.

Since double adoptions provide for perfect compatibility of S_1 to S_b, it does not matter for agents of *pop*1 whether agents of *pop*2 adopt S_b or S_1. Moreover, the fitness of S_2 for agents of *pop*2 is not affected by how many agents of *pop*2 adopt S_b. Thus, in our framework, there is neither a negative external effect nor any collective action problem produced by the availability of double adoptions. Nonetheless, through a switch from S_2 to S_b, agents from *pop*2 confer a positive externality on agents from *pop*1. Thus, private incentives to switch to S_b tend to be still too low.

4. Mandatory or Voluntary Standards?

4.1 Introduction

Our analysis in section 3 suggests that globalization does not produce enough global standards. Even in cases where harmonization of national standards actually occurs, it does so too late from a societal point of view. Double adoptions or converters, if available, may help yet not eliminate the problem. Moreover, the "wrong" global standard may prevail. Of course, in face of potentially severe side-effects, the results of our model alone cannot justify policy intervention. However, the WTO Agreement on Technical Barriers to Trade concluded within the framework of the Uruguay Round in 1993, as well as considerable efforts on the EU-level are evidence that policy-makers do actually consider more harmonization of national standards as a desirable goal.[87]

Therefore, it may be worth to check whether we can use our model to derive recommendations onto the problem *how* to optimally achieve harmonization of national standards.[88] One problem within this context is the legal status of a (harmonized) global standard: should its adoption be mandatory or voluntary (ore something in between)? We offer a simple line of arguments on this question. Assume we want that people from *pop2* give up their domestic standard S_2 and switch to S_1, the domestic standard of the *pop1*. The obvious way to achieve this is to make S_1 a mandatory standard, *i.e.* to impose a (expected) sanction on the non-adoption of S_1. However, even a voluntary "formal" standard might suffice to implement harmonization. Consider Figure 10 (section 3.3.4) again. One can see that not *all pop2* has to be induced to switch to S_1. Once a sufficient share of agents has switched to S_1, the fitness of S_1 has become larger even for agents from *pop2*, and so the rest of them will follow to S_1.[89] Thus, whether we can use a voluntary global standard for achieving harmonization depends on whether it produces the required collective switch to S_1. It seems to be a natural assumption that such a standard can in fact do so if the required collective switch is sufficiently small, *i.e.* if the "stability of variety" is sufficiently low.

Another factor that determines the optimal choice between voluntary and mandatory standards is the costs of these instruments. Since a mandatory standard has to pass the legislative procedures and may require enforcement costs it seems likely that it is more costly to estab-

---

[87] See Sykes (1995), p. 63-85, for details on the WTO agreement and section 4.4 below for the EU harmonization policy.

[88] We restrict our analysis in this section to the case where standards are mutually exclusive, *i.e.* double adoptions are not available.

[89] Of course, harmonization is self-reinforcing as long as "spontaneous" deviations are not too large.

lish such a standard. Thus, we can derive a simple recommendation: if at a particular harmonization project the stability of variety is low, a formal standard should be voluntary; if the stability of variety is large, a mandatory standard should be used.

We consider the effects of voluntary and mandatory standards within our model in more detail in section 4.2. Since it plays a crucial role for the optimal choice of the legal status for a standard, the stability of variety is analyzed in more depth in section 4.3. As a next step, it makes sense to investigate the harmonization policy in a particular area with regards to its use of voluntary and mandatory standards. In section 4.4, we consider the harmonization of "technical standards" within the EU. Unfortunately, we do not "test" whether the harmonization policy is designed according to our simple theory, *i.e.* whether the choice of legal status of standards depends on the stability of variety. Instead, we show how technical harmonization policy within the EU makes use of mandatory and voluntary standards and demonstrate that this practice, through shifting responsibility to the EU's "official" standardization bodies, entails the risk of excessive harmonization.

4.2   The Effects of Mandatory and Voluntary Standards

Given populations are stuck in a stable variety state, any intervention by an "authorized body"[90] that aims at inducing harmonization must (in the end) cause all individuals from one population (say *pop j*) to adopt the standard of the other population (say *pop i*).[91] Thus, the fitness of the strategies S_i and/or S_j with respect to *pop j* have to be manipulated, *i.e.*, any successful intervention must produce fitness values such that $f_{ji} \geq f_{jj}$. Recall that the fitness of the strategies depends only on two factors: the probability of being matched with someone using the same standard and the respective interaction payoff (given in the matrix in Figure 8). Thus, two kinds of instruments are available: collective switches that change the probabilities and manipulations of the underlying interaction payoffs.

4.2.1   Inducing a Collective Switch through Voluntary Standards

In order to produce harmonization with S_i an authorized body may manipulate p_{ji}. In evolutionary terminology, the body must produce a sufficiently large "invasion of mutants" such that $f_{ji} \geq f_{jj}$. Thus, not all *pop j* has to be induced to collectively switch. Once enough agents from *pop j* have switched to S_i, the rest will follow. The minimum invasion that produces

---

[90]   For the clearness of terminology: we call an "authorized body" a body that is empowered to set rules.
[91]   Of course, one may also choose a harmonization candidate that has not yet been adopted. We exclude this possibility here.

$f_{ji} \geq f_{jj}$ corresponds to the stability of variety, also known as its "basin of attraction".[92] Building upon a corresponding "static" formulation, introduced by Peters (1997, 1998), we define the minimum invasion of mutants required to upset variety and implement harmonization with S_i as

(DIII) $\varepsilon^{CM}(V \rightarrow S_i) := \varepsilon \mid f_{ij}(\sigma_j - \varepsilon) = f_{ji}(\sigma_j - \varepsilon); \; i,j \in \{1, 2\}; \; i \neq j$ [93]

See Figure 14 (resembling Figure 10) for illustration. Recall that $\pi = \sigma$ corresponds to variety. The chosen parameters imply that harmonization with S_1 yields maximum welfare. To induce harmonization with S_1, not all *pop2* needs to switch but rather just marginally more than 0.1 agents. Once just above 0.8 agents adopt S_1, f_{21} becomes larger than f_{22}, and so the evolutionary process eventually leads to harmonization with S_1.

Figure 14
Stability of variety

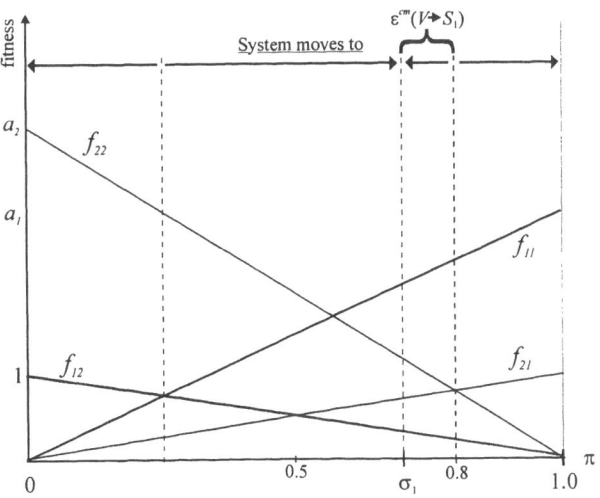

Exogenous parameters: $a_1 = 3$, $a_2 = 4$, $\sigma_1 = 0.7$, $\omega = 1$.

---

[92]   See, *e.g.*, Mailath (1998).

[93] σ_j denotes the size of *pop j*. Note that this stability definition requires monotonicity of f_{ji} (which applies in our setting).

Young (1993) and Kandori, Mailath and Rob (1993), for example, assume that mutants arise stochastically and hence might produce an invasion large enough to generate a shift from one stable equilibrium to another. Apart from the risk that such a process might yield (temporary) harmonization with an inefficient standard, it might – depending on the specific mutation rate – take too long before variety is left in the first place. Hence, even if mutants do pop up from time to time, policy intervention may be perceived as desirable.

Holler and Peters (1999) and Holler and Wickström (1999) suggest that a "scandal" might generate a collective switch. Holler and Peters argue that, unlike animals, humans are able to deliberately change their behavior. However, costs of coordination (as it requires costly communication) limit the extent of potential collective changes. If a population is stuck with an inferior social norm, a scandal might help. A celebrity deviating from the established norm might receive high publicity, as this is published in the press and on TV and, hence, can serve as a means of communication, implementing a superior norm. As an example of such scandals, Holler and Wickström propose that the sexual revolution in the film industry during the 1960's might have gone through such a process. "The ice was broken by a few outstanding directors like Felline ('La dolce vita') and Bergman ('Tystnaden'), and the bandwagon started rolling" (Holler and Wickström 1999, p. 104).

How can an authorized body *deliberately* produce a sufficiently large invasion of mutants? Cooter (1998) proposes that a law – *even if not enforced* – may produce a collective switch. Similar to Holler and Wickström and Holler and Peters, Cooter argues that a law can serve as a means of coordination: "... think of law as solving a problem of collective action ... In an effective democracy, citizens respect the law and feel obligated to it. Law making is a collective decision that could induce the coordination required to change to a Pareto superior equilibrium" (Cooter 1998, p. 13). Cooter suggests that the prohibition of smoking in US airports was based on such a mechanism. "Most people began to obey this law as soon as they became aware of this law. For the small groups of lawbreakers, rude remarks by citizens and other informal punishments deter without state coercion" (Cooter 1998, p. 13).[94]

It would be consistent with the above arguments of Holler and Wickström (1999), Holler and Peters (1999) and Cooter (1998) to claim that not only scandals and laws (whether or not enforced) may induce a collective switch. Due to their awareness in public and/or in respec-

---

[94] What might have happened is this. There are two equilibria: all smokers smoke and no smoker smokes in airports. If all smokers smoke, non-smokers put up with smoking, since it is too much of a hassle to entreat so many smokers. If only a few smokers smoke, non-smokers find it beneficial to apply social sanctions on smokers. The introduction of the law has pressed the "reset button": imagine someone who likes to smoke enters an airport for the first time after the law has been enacted. Probably he will hesitate at first. Then, if he sees many people smoking he would smoke as well. If hardly anyone is smoking he would – anticipating informal punishments – also refrain from doing so.

tive industries and their "official" character, also formal standards issued by recognized standardization bodies such as DIN, CEN and ISO and similar "rules" such as guidelines from self-regulatory bodies (*e.g.* the GAAP) may have this effect.

4.2.2 Changing Interaction Payoffs through Mandatory Standards

Instead or in addition to induce a collective switch an authorized body could change the *interaction* payoffs. For example, to produce harmonization with S_i, it could make the adoption of S_i mandatory and impose a sanction on the adoption of S_j. Fitness of S_j results as

(E14) $f_{ij} = p_{ij}(a_j\text{-}s) - p_{ji}s$

where s denotes the (average) sanction to be paid when interacting with S_j.

In Figure 15, the arrows located at the top show where the system moves, depending on π, the fraction of S_1-users. Sanctioning S_2 in *pop2* shifts f_{22} downwards. The (average) sanction chosen in the figure is 1.67. Situated at variety, this sanction just suffices to render S_1 fitter even for agents from *pop2*. This triggers the evolutionary process, making the populations end up with harmonization with S_1.[95]

It would be consistent with the arguments of the previous subsection that also the introduction of a mandatory standard produces a collective switch – in addition to the just described effects of directly reducing the interaction payoff. In fact, due to the threat of a sanction, the public awareness of mandatory standards is likely to be even larger than of voluntary standards.

An example for a mandatory standard is the British Weights and Measurements Act of 1985 and the Weights and Measures (Metrication Amendments) Regulations 2001 (came into force on 8th February 2001). These rules, which are supplemented by sanctions, prescribe that British vendors have to sign their goods according to the metric system.

---

[95] Similarly, a subsidy to S_1 could promote harmonization with S_1.

Figure 15

Manipulation of S_2's fitness by sanction

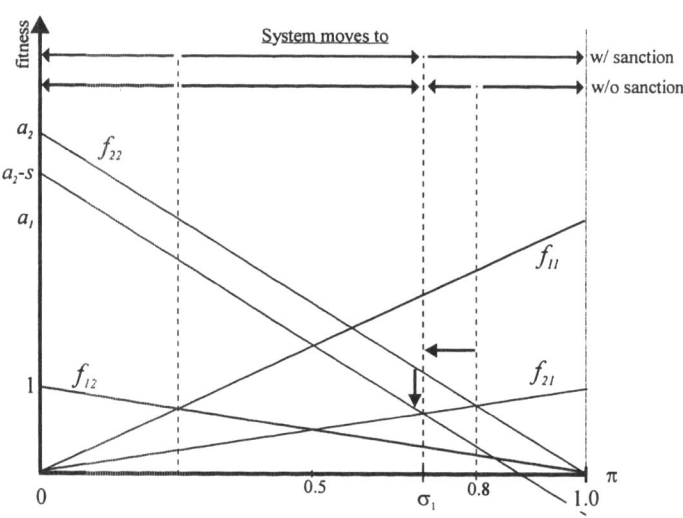

Exogenous parameters: $a_1 = 3$, $a_2 = 4$, s = 1.67, $\sigma_1 = 0.7$, $\omega = 1$.

4.2.3 Which Kind of Standard is optimal?

We have argued that in order to implement harmonization, an authorized body can use mandatory or voluntary standards. As harmonization is the only purpose considered here, an authorized body should apply the measure associated with the least costs. It seems reasonable to assume that mandatory standards are more expensive than voluntary ones. This is obvious because a mandatory standard has to pass the legislative procedures and effective sanctions require enforcement activities, administration and redistribution costs. Furthermore, a mandatory standard might impose more inflexibility on agents than is necessary to implement harmonization.[96]

On the other hand, the collective switch induced by a voluntary standard may not always be large enough to implement harmonization. In fact, it seems intuitive that such a switch is very likely to be smaller than 1. First, agents must actually receive the information that the rule has been introduced. Obviously, it is likely that not each agent would be informed. Second, agents

---

[96]   Since each population is homogenous, this would not occur within our model, though.

may form different expectations with regards to how many people will obey the rule. As a result, the larger the stability of variety, the more likely it is that such a rule fails to induce harmonization. Attempts to replace the imperial system by the metric system in the USA and UK on a voluntary basis is a good example for a case where collective actions have not been large enough. Yet, all these attempts have failed.

Thus, our ideas suggest a simple theory for optimal harmonization policy: depending on the particular harmonization project, if the stability of variety is "low", voluntary standards are optimal; if the stability is "high", mandatory standards are optimal. Of course, there are statuses in between; think, *e.g.*, of default rules. For simplicity, we only consider the extreme cases. The analysis is easily extended so that it includes "intermediate" statuses.

4.3 The Stability of Variety

Since the foregoing discussion suggests that the optimal intervention may depend on the stability of variety, we now look at this issue in more detail: on what factors does the stability of variety depend, in turn?

First of all, note that variety's stability depends on which standard is targeted for harmonization. Since the populations regularly differ in size and preferences, variety's stabilities $\varepsilon^{CM}(V \to S_2)$ and $\varepsilon^{CM}(V \to S_1)$ typically differ. For example see Figure 14 again, in which $\varepsilon^{CM}(V \to S_2)$ is much greater than $\varepsilon^{CM}(V \to S_1)$.

4.3.1 Formal Expressions

Plugging the expressions for the fitness values from (E2) into (DIII) and solving for $\varepsilon^{CM}(V \to S_i)$ gives us:

(E15a) $$\varepsilon^{CM}(V \to S_i) = \left[\frac{a_j}{\omega(1 + a_j)} - \sigma_i \right] \left[\frac{\omega(1 - \sigma_i)}{1 - \omega\sigma_i} \right]$$

For $\omega = 1$, this simplifies as:

(E15b) $$\varepsilon^{CM}(V \to S_i) = \frac{a_j}{1 + a_j} - \sigma_i$$

where σ_j is the size of *pop j*. Recall, at variety, σ_i is the share of agents who adopt S_i and σ_j is equal to the share of agents that play S_j .[97]

For intuition on these conditions see that in (E15b), the term $(1+a_j)$ is the difference of the slope of the fitness function f_{jj} and f_{ji} within the relevant area. For $\omega < 1$, the corresponding term is $[(1-\omega\sigma)/(1-\sigma)][1+a_j]$. With each "marginal switch" from S_j to S_i the fitness values of the two strategies with respect to agents of *pop j* come closer to each other by an amount equal to this term. With ε^{CM} deviations f_{jj} and f_{ji} are equalized. Any additional switch takes the system to harmonization with S_i.

In (E15a), the numerator of the second factor corresponds to the probability that an agent of *pop j* interacts with a member of his own population on the world market. The denominator is the overall probability of a member of *pop j* interacting with someone of his own population. With increasing ω, these probabilities come closer to each other; with $\omega = 1$, these probabilities are equalized, and (E15a) turns into (E15b).

As we can see from (E15a), $\varepsilon^{CM}(V{\rightarrow}S_i)$ only depends on the relative sizes of the populations, the preferences of the population whose members switch to the alien standard (a_j), and on the integration level (ω). Notice that the preferences of *pop i*, a_i, do not enter $\varepsilon^{CM}(V{\rightarrow}S_i)$. This makes sense, as only members of *pop j* switch and a_i neither enters f_{ji} nor f_{jj}.

4.3.2 Variety's Stability and the Integration Level

Figure 16 illustrates that the lower ω the larger is $\varepsilon^{CM}(V{\rightarrow}S_i)$. In addition to the fitness functions for $\omega = 1$, Figure 16 includes the fitness functions for members of *pop2* for $\omega = 0$ and for ω', $(0<\omega'<1)$.[98] Recall, with $\omega = 0$, interactions take place only between members of the same population. Thus, at variety, coordination with the preferred standard is achieved in any interaction. The larger ω the smaller is the fraction of coordinated interactions conducted at variety. Thus f_{22} and f_{11} shift downwards when ω intensifies.

On the other hand, the "leverage" of switches is larger if ω is low. f_{22} decreases even more with each agent from *pop2* switching to S_1 because the probability to meet an agent from *pop2* is larger if ω is low. Moreover, f_{21} increases more with every agent of *pop2* switching to S_1. However, for any $0<\pi<1$, a decrease in ω always shifts f_{22} upwards and f_{21} downwards. Thus,

---

[97]   Our stability expressions build upon the fact that, in our two-population setup, the collective switch that produces a shift from variety to harmonization is homogenous because at variety all agents use their preferred standard. In order to achieve harmonization with S_i, only agents from *pop j* switch (to S_i). Also note that if $\varepsilon^{CM}(V{\rightarrow}S_i) = 0$ variety is not ESS-stable; if $\varepsilon^{CM}(V{\rightarrow}S_i)$ is negative variety is not an equilibrium.

[98]   Recall that our figures use the simplifications regarding the interrelation of π and the strategy profile (φ_1, φ_2) as described in section 3.3.4.

the intersection of f_{22} with f_{21} shifts to the right when ω decreases. Hence the stability of variety decreases with increasing integration.[99]

As a result, globalization strengthens the need for harmonization but it also makes it easier to achieve harmonization. With low integration, the benefits of harmonization, if positive at all, are low and harmonization is likely to be costly to reach. Thus, harmonization may be a matter of timing. Even if the benefits from harmonization are greater than the costs to be incurred in achieving it, it might in some cases be better to wait until globalization has further advanced and soft low-cost measures can be applied.

Figure 16
Variety's stability and integration

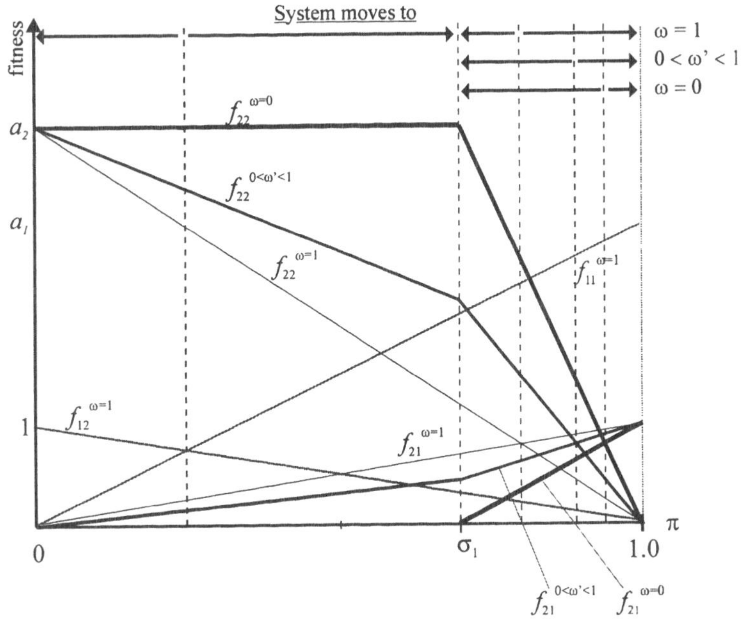

Exogenous parameters: $a_1 = 3$, $a_2 = 4$, $\sigma_1 = 0.7$, $\omega = 1$ ($\omega = 0$, $0 < \omega = \omega' < 1$).

4.3.3 *Variety's Stability and the Agents' Preferences*

The larger a_j the greater is $\varepsilon^{CM}(V \rightarrow S_i)$. With low level of ω, the effect of a_j on $\varepsilon^{CM}(V \rightarrow S_i)$ is even stronger. Recall that all agents of *pop i* adopt S_i. So, in relevant states, the smaller ω is,

---

[99] Of course, an analogous reasoning applies to $\varepsilon^{CM}(V \rightarrow S_2)$.

the larger is the probability of meeting an agent playing S_j for agents of *pop j*. Thus, the impact of an increase in a_j intensifies when ω decreases.

One can conclude that even if it is necessary to supplement a standard with costly sanctions (or subsidies), it might in some cases pay off to wait for further integration. Lower sanctions can be set and/or enforcement activities can be reduced.

4.3.4 Variety's Stability and the Populations' Sizes

If the integration among the populations is perfect ($\omega = 1$), the stability $\varepsilon^{CM}(V{\rightarrow}S_i)$ increases with σ_j by a rate of 1. That is, each additional member of *pop i* (which is also one agent less in *pop j*) reduces the collective switch that is required to reach S_i by exactly one additional agent. This is due to the fact that (with $\omega=1$) the fitness functions are independent of the relative sizes of the populations, which, in turn, is a consequence of our assumption that interaction payoffs only depend on the adopted strategies, not on the affiliation of the opponents.

If the integration is imperfect ($\omega < 1$), $\varepsilon^{CM}(V{\rightarrow}S_i)$ increases with σ_j by a positive rate, as well. However, this rate is smaller than 1. To see why, remember that $\omega < 1$ implies that the fitness functions do depend on the populations' sizes. An agent of *pop i* meets an agent of *pop j* with a different probability than an agent from *pop j* does. The expected payoff of a strategy when interacting on the domestic market does *not* depend on the size of the domestic market because at variety, everyone in that market plays the same strategy. Thus, $\omega < 1$ implies that the domestic standards' fitness is less dependent on the populations' relative sizes. Furthermore, the fitness of the alien standard increases less in σ_i because the probability that an agent from pop *j* is matched with an agent from *pop i* is smaller than σ_i (while it is σ_i with $\omega = 1$).

For illustration see Figure 17, which includes f_{22} and f_{21} for $\omega = 0$. If σ_1 decreases from σ_1' to σ_1'', the variety equilibrium shifts to the left by the same amount. Thus, for $\omega = 1$, $\varepsilon^{CM}(V{\rightarrow}S_1)$ increases by exactly this amount. For $\omega < 1$, however, there is an additional effect. While f_{22} rotates counter-clockwise, f_{21} rotates clockwise, so their intersection shifts to the left, as well. This leads to a reduction of the first effect on $\varepsilon^{CM}(V{\rightarrow}S_1)$.

Figure 17

Variety's stability and populations' sizes

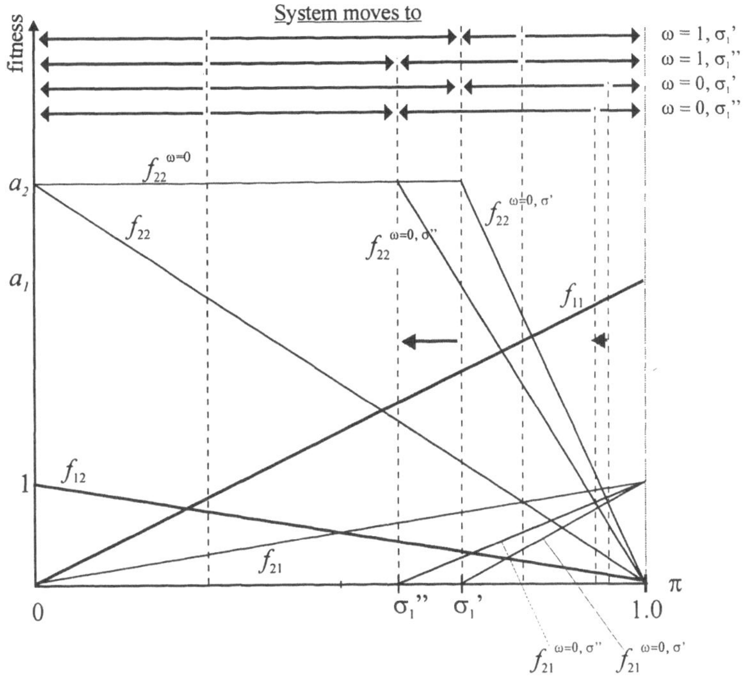

Exogenous parameters: $a_1 = 3$, $a_2 = 4$, $\sigma_1' = 0.7$, $\sigma_1'' = 0.6$, $\omega = 1$ ($\omega = 0$).

4.3.5 Variety's Stability and Efficiency of the Global Standards

A comparison of (E3)–(E5) with (E15a) indicates that the stability of variety with respect to the efficient global standard might very well be larger than the stability with respect to the inefficient one. This result is due to the fact that the preferences of *pop i* determine the welfare level of harmonization with S_i while they are irrelevant for the size of the collective switch needed to achieve harmonization with S_i. See Figure 18 for illustration. With $a_1 = 3$, welfare is maximized when harmonized with S_1 ($W_{S1} > W_{S2}$). If a_1 shifts down to 2, S_2 becomes the superior harmonization candidate ($W_{S2} > W_{S1}$). $\varepsilon^{CM}(V \rightarrow S_1)$, however, remains unchanged. While $\varepsilon^{CM}(V \rightarrow S_2)$ becomes smaller, it is still greater than $\varepsilon^{CM}(V \rightarrow S_1)$. In the case depicted in Figure 18, the collective switch needed to achieve harmonization with the efficient S_2 is actually more than twice as large as the one required to reach harmonization with S_1.

Figure 18

Variety's stability and efficiency of harmonization candidates

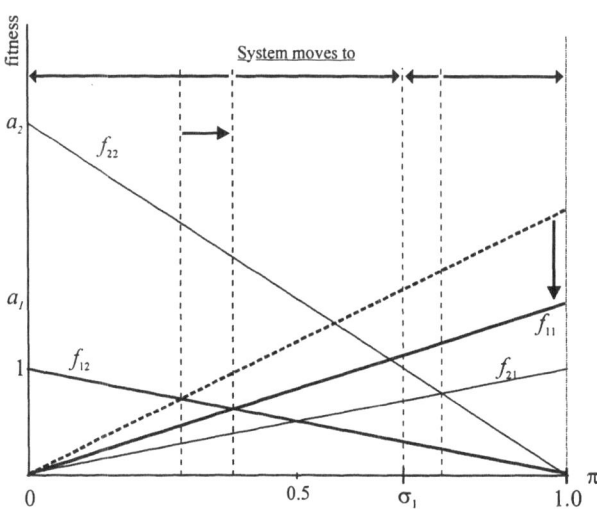

Exogenous parameters: $a_1 = 2$, $a_2 = 4$, $\sigma_1 = 0.7$, $\omega = 1$.

Thus the "statically" efficient global standard might be the one that is harder to reach.[100] In such a case, a benevolent authorized body trades off the welfare loss due to harmonization with a sub-optimal standard with the costs saved to achieve harmonization. Note, however, that this might be a risky choice. Once harmonization with the "statically" inferior harmonization candidate is implemented, it is even more difficult to obtain harmonization with the superior global standard.

4.4   Technical Harmonization within the EU

We now apply our analysis on technical harmonization within the EU. Through its 1985er resolution the EU Council established the "New Approach on Technical Harmonization and Standards".[101] According to this concept, large parts of the European technical harmonization work are delegated to the EU's "official", however non-governmental, standardization bodies (ESBs), namely the CEN, the CENELEC and the ETSI. These bodies operate in different

---

[100]   Of course, the "statically" efficient global standard might become the inefficient one after taking into account the costs of producing harmonization.

[101]   See Official Journal of the European Communities (OJ), C 136 of 4.6.85, 1-9.

fields. The CENELEC is responsible for electrotechnique. The ETSI is specialized in the tele-communication field and the CEN covers the remaining areas. The national standardization bodies (NSBs) like the German DIN and the VDE, the French AFNOR and the UTE, etc., make up the "members" of the CEN and the CENELEC, respectively.[102] Note that neither the ESBs nor most of the NSBs are governmental bodies. The CEN, for example, is a non-profit organization under Belgian law; the DIN is a "eingetragener Verein". The French AFNOR, one of the few exceptions, is a government agency.

Besides developing formal standards in new areas, these bodies aim at harmonizing national standards. They do not only deal with compatibility standards (as considered here) but also with technical rules concerning safety and health, quality and environment for products as well as production processes: "The standards cover products, systems, and services and pro-mote rationalization and quality assurance in the fields of industry, technology, science and administration. They ensure interworking and interchangeability of products and systems, rational order and communication between market partners."[103]

4.4.1 The Process of Technical Harmonization in Europe

Figure 19 describes the process of "official" technical standardization and harmonization in the EU. According to the New Approach, European Harmonization Directives formulate "es-sential requirements" to which products must conform in order to comply with national regu-lations within the EU (and so enjoy free movement). These directives are supposed to elimi-nate barriers of trade created by disparity of national regulation concerning the health and safety of citizens and consumers, environmental protection or other protected common goods.[104]

Based on a particular European Harmonization Directive, the European Commission may "mandate" the development of standards to the appropriate ESB, in order to specify character-istics for products that meet the respective essential requirements. Note that the adoption of these standards is supposed to remain voluntary. Firms are free to otherwise prove conformity to the essential requirements (if they want to enjoy free movement of their goods within the EU). There are exceptions, though. For example, the Directive relating to Telecommunica-tions Terminal Equipment and Satellite allows in respect of certain essential requirements (electromagnetic compatibility, safety of the public telecommunication infrastructure and cer-

---

[102] ETSI is organized somewhat differently.

[103] See clause 1.3 of the Joint CEN/CENELEC/ETSI statement "Basic principles and organization of European standards work".

[104] See OJ C 136 of 4.6.85, 2-3.

tain network aspects) that harmonized standards are transformed into common technical regulation, with which compliance is mandatory.[105] (This example also shows that standards, even
if developed to meet health and safety requirements, may affect issues of compatibility.)

Important is, however, that most European technical standards are initiated by the ESBs
and/or NSBs themselves. For example, in 1999, only about a quarter of the "ratified" CEN
standards were connected to "New Approach Directives".[106]

Figure 19

The process of technical harmonization in Europe

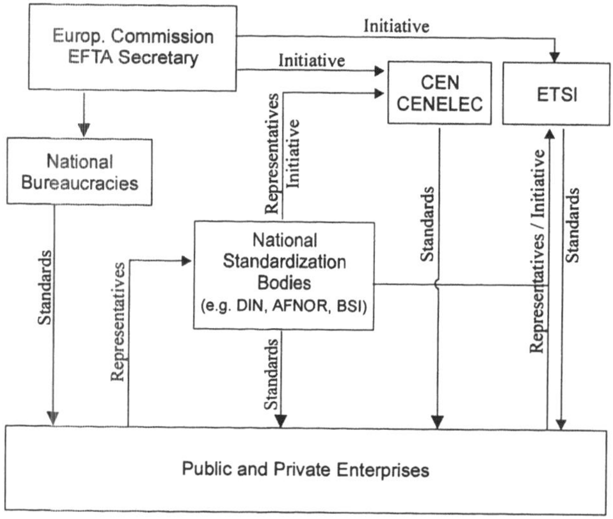

Source: Goerke and Holler (1998), p. 98.[107]

The actual development of standards usually takes place in the "Technical Committees" of
an ESB, which consists of representatives of each NSB. For each European Technical Committee each NSB usually sets up corresponding national committees, which are open to all
interested parties. In these national committees, the national positions are worked out and

---

[105] See Articles 5 lit. c-g, 7(2) and 29 of the Directive 981/13/EC (OJ L 74 of 12.3.98, 1-26). See also Schepel
       and Falke (2000), p. 24.
[106] See Falke and Schepel (2000).
[107] For a more detailed schedule see Schepel and Falke (2000), p. 146-7.

supposed to flow into the discussion within the European Technical Committee.[108] Finally, after enquiry and evaluation of public comment, a vote is taken.[109] If successfully passed, the standard is "ratified", which then makes it obligatory for all NSBs to transpose the standard into a national standard.

4.4.2 How European Standards Produce Harmonization

Generating a collective switch through voluntary standards

As noted in 4.4.1, most harmonization initiatives are launched by the ESBs or NSBs themselves. These "official" standards are not related to Harmonization Directives. Still, it appears consistent with our argumentation in section 4.2.1 that such European official standards have some "harmonization power", since it is likely that they generate a collective switch. First of all, the ESBs and their members, the NSBs, are well known and widely recognized. This is due their "official" and "exclusive" status, which they have been granted by the EU Council and the national governments, respectively.[110] Their standards tend to receive a high degree of awareness in the respective industries. Thus, consistent with the above arguments of Holler and Wickström (1999), Holler and Peters (1999), it is likely that these official standards serve as a means of (implicit) communication. Second, the ESBs and NSBs give a platform for actual (explicit) communication. As described in section 4.4.1, the national Technical Committees invite all interested parties to discuss and coordinate their position.

Moreover, as Clark (2000) points out, official standards can create "legitimate expectations".[111] People often actually think that standards are legally binding. Such expectations may receive further support from the fact that some standards are in fact made mandatory by law (see below). Thus, paralleling Cooter's argumentation, individuals may – at least initially – feel obliged to use these standards.[112] Several non-formal interviews this study's author conducted with people from the construction industry confirm Clark's hypothesis.

---

[108] See Woeckener (1997), p. 392.

[109] The voting weights in the ESBs are allocated similar to those in the EU Council. Approval requires qualified majority. See section 6 for details.

[110] See section 4.4.1 above for the ESBs. Most NSBs have signed a contract with their national government, which recognized the NSB as the "competent Standard body" or similar. See, for example, the "Normenvertrag" between the German government and the DIN signed in 1975.

[111] See Clark (2000), pp. 482-3.

[112] See section 4.2.1.

Evidence for the presumption that the ESB's standards' do have the power to initiate collective switches is certainly given by the fact that many of their purely voluntary standards do establish as de facto standards.

Manipulate interaction payoffs through mandatory standards

Although heavily emphasized that the adoption of official standards is voluntary, non-adoption may well be sanctioned via the law and therefore – beside collective action power generated by establishment as ESB standards – support harmonization.[113] There are several ways how legislations within the EU indeed do so.

As already mentioned in the previous section 4.4.1, those official European standards that are mandated and approved by the European Commission lead to the presumption of conformity to national European mandatory regulations (implementing European Directives). Thus, although not legally binding, the adoption of such harmonized official standards may be "necessary" in fields where legal uncertainty with respect to national regulation is relevant. Of course, in such fields, this will, within our framework, then be effectively equivalent to a sanction of non-adoption of the official standard.[114]

In addition, national laws often use "blanket clauses" such as "state of the art", "acknowledged rules of technology" and "good practice" among others. These blanket clauses are widely used for safety and quality regulations, though. Although not automatically, the adoption of the official respective standards often leads to presumption of complying with these laws. In a 1988 judgment, the German *Bundesgerichtshof* states:

> "Even if standards are not norms which bind third parties in the sense of formal legis-
> lation, but rather recommendations of voluntary application of the DIN (...), they do reflect
> the applicable acknowledged rules of technology in the circles concerned and are therefore
> particularly suited to determine the safety required according to the prevailing views in this
> sector"[115].

The *Bundesverwaltungsgericht* in its 1997er statement bounds this view:

> "It is true that the federal legislator refers to 'acknowledged rule of technology'. These
> rules, however, do not constitute legal norms. The DIN has therefore no legislative power.
> It is an association, which as made its statutory objective to produce standards in the gen-
> eral interest through the collective efforts of interested circles with a view to rationaliza-
> tion, quality assurance, safety and compatibility. They have legal relevance not because of

---

[113]   One exception is France where official national standards *as such* are legally binding to some extent.
[114]   Or, depending on the point of view, to a subsidy of the adoption of the official standard.
[115]   BGHZ 103 of 1.3.1988. Translation by Schepel and Falke (2000), p. 206.

their autonomous validity but only in as far as they fulfil the constitutive conditions of ac-knowledged rules of technology, which the legislator is concerned with as such. When the legislator refers to them, they have a normative function in the sense they substantiate the material legal norm"[116].

Nevertheless, the adoption of official standards does often lead to presumption of confor-mity to the respective law.[117] Thus, in a way akin to the conformity presumption to essential requirements, such practice is likely to work similar to a sanction of non-adoption of stan-dards.[118]

In addition to these "semi-mandatory" standards, it is a common practice within most Euro-pean legislations to explicitly declare the application of some European "official" standards mandatory. These references can be rather general and even "undated" such that revisions of the standards do not even require a modification of the law. For example, the Belgian law on mobile telephony networks states that "the system put in place by the operator must conform to the relevant ETSI standards"[119]. According to the German *Verordnung zur Anwendung von Normen für voll digitale Fernsehdienste*", must transformation-systems comply with se-veral ETSI norms.[120] Another example is § 11 IV *Bay. SchiffahrtsO* of 19.6.1968, which states that "Electric Equipment must comply to the rules of the VDE."[121]

Such dynamic, "undated" reference is widely considered to be unconstitutional and un-democratic (at least in Germany, Austria and Switzerland).[122] It is claimed that it is a hidden delegation of legislative power to private organizations; such practice offends the principles of democracy and the fundaments of constitutional state, such as the doctrine of assertiveness and legal clarity, the doctrine of separation of powers and the principle of publication.[123] On the other hand, however, this practice can be viewed as an efficient means of regulation

---

[116] Translation by Schepel and Falke (2000), p. 188, from Neue Zeitschrift für das Verwaltungsrecht (1997), Rechtsprechungs-Report, p. 241.

[117] See Ebert-Kern (1994), p. 43.

[118] The German *Energiewirtschaftsgesetz*, for example, goes even one step further: it prescribes the application of "acknowledged rules of technology" for construction and run of energy plants and even explicitly states that this is presumed if the respective VDE-standards are used (see § 16 I *Energiewirtschaftsgesetz*; see also Falke (2000), p. 269).

[119] Article 6, *Arrêté Royal relatif à l'établissement et à l'exploitation de réseaux de mobilophonie GSM*, MB 8/4/1995, p. 9048. See also Schepel and Falke (2000).

[120] See *Verordnung zur Anwendung von Normen für voll digitale Fernsehdienste* of 4.2.1999 in Bundesgesetz-blatt (BGBl) I S. 85.

[121] The translations have been done by this study's author.

[122] See Marburger (1987), p. 835. See also Schepel and Falke (2000), p. 189.

[123] See Falke (2000), p. 251.

whose advantages might be worth being traded off.[124] It is not the right place to contribute to this discussion. Whether or not such references are admissible, it appears most likely that they enter the payoffs individuals base their decisions on.[125] Thus, such references do sanction the non-adoption of respective official standards.

There is another way that lawmakers use to promote official standards: the Public and Utilities Procurement Directives render the adoption of harmonized official standards mandatory in public procurements.[126] For example, Article 10 of the Public Works Directive states:

> "Without prejudice to the legally binding national technical rule and insofar as these are compatible with Community law, the technical specifications shall be defined by the contracting authorities by reference to national standards implementing European standards, or by reference to European technical approvals or by reference to common technical specification."

There are exceptions to this, though. In essence, public procurement entities are allowed to deviate from this rule if standards do not provide for conformity assessment, if (however only on a transitional basis) standards are incompatible with material already in use, or if the project is of genuinely innovative nature where use of existing standards would be inappropriate.[127]

How can we integrate the procurement entities' obligations to refer to (harmonized) standards into our evolutionary framework? If public procurements, as for instance in the case of the construction industry, make up a relevant share of the industry's total demand, we could integrate special player with which agents are matched with a larger probability than with regular individuals. Consequently, a switch of the public procurement entities from, say, S_j to S_i, results in a (non-marginal) increase in p_{ii} and p_{ji}, according to the governments' shares of total industry demand.[128] Thus, within our framework, the effect of making harmonized standards mandatory in public procurements is similar to a collective switch.

In Germany's civil construction industry, there is another reason why official standards have a great deal of impact. The DIN closely cooperates with the *Deutscher Verdingungsausschuß für Bauleistungen* which is responsible for development and maintenance of the VOB (*Verdingungsordnung für Bauleistungen*). VOB is a kind of standard contract. In its part C, it

---

[124] See, *e.g.*, Schepel and Falke (2000), p. 187.

[125] After all, these laws (even recent ones) do exist. Thus, it is most likely that infringement of such a law is associated with an expected sanction, whether through penalty, damage payments or court costs.

[126] See Council Directives: 92/50/EEC, 93/36/EEC, 93/37/EEC, and 93/38/EEC.

[127] See also Schepel and Falke (2000), p. 207.

[128] Of course, the extent by which these probabilities increase also depends on the level of integration as well as the relative sizes of the populations.

contains relevant DIN standards. Although only part A is mandatory for public procurements, de facto, the entire VOB (subject to contract specific adjustments) is of vital importance to the German construction industry and is used for almost any construction project.[129]

4.4.3 Results

There are several results one can derive from the above analysis. Our simple theory of optimal intervention for harmonization described in sections 4.2 and 4.3 suggests that voluntary global standards are optimal if variety's stability is low and mandatory standards are optimal if variety's stability is high. We have found that European harmonization policy makes in fact use of both kinds of instruments. In most cases, it makes use of voluntary standards, *i.e.* it solely relies on the collection switch effect of European official standards. Through making the ESBs and NSBs both "official" and "exclusive", it strengthens the "collective switch power" of such standards.[130]

Some standards, however, are supplemented by sanctions based on mandatory or "semi-mandatory" legislation. If mandated by the European Commission, ESB's standards, through the threat of non-compliance with mandatory regulation within the EU, may, in effect, work like mandatory standards. Besides, other technical standards – at least for use in specific applications – are directly or indirectly declared mandatory. Moreover, harmonized official standards are (partly) mandatory in public procurements.

One would have to test whether such mandatory standards are in fact only applied if voluntary ESB standards are not sufficient to overcome variety's stability. There remain some doubts. For example, through making European standards mandatory in public procurements, the harmonization power of official standards heavily depends on the government's share of total industry demand. At least the study's author does not see any reason why variety's stability should corresponds to this share.

The European harmonization practice may entail another problem, which concerns the timing of harmonization. This is based on two results. First, the desirability of harmonization is an increasing function of the integration level (see section 3). Unless integration among European nations in the considered industry has not grown beyond some threshold (ω^{CR}), variety remains the efficient state. Second, as shown in section 4.3, costs of necessary measures to produce harmonization tend to decrease with increasing integration. A benevolent harmonization body would account for both issues. He would only consider intervening when the inte-

---

[129] This, in turn, may be the result of network effects or path dependence.

[130] See, *e.g.*, Directive 98/34/EC giving CEN exclusive status.

gration has passed ω^{CR} in the first place and would then trade off the benefits of early harmonization against the costs of necessary measures.

"It appears, however, that there is a bureaucratic component involved in the objective functions of the [ESBs] which is due to the staff employed by them" (Goerke and Holler, p. 101). The ESBs and NSBs, despite their power, do neither participate in the harmonization expenses nor (hardly) in the returns from harmonization. For instance, almost half of the CEN's budget is financed by governmental funds (see Figure 20a). Furthermore, the NSBs control the ESBs through exclusive membership. The NSBs, in turn, fund most of their work through sale of standards (see Figure 20b). Earnings certainly increase through harmonization of standards. Thus, budget maximizing NSB and ESB bureaucrats would have excessive incentives for harmonization.[131,132] On the other hand, such bodies' "harmonization power" is strong. Not only do they apparently have the power to induce collective switches. As shown in section 4.4.2, some of their standards directly or indirectly enter into mandatory laws. Thus, one may fear that the ESBs' produce too much harmonization in Europe.

Unfortunately, our stability analysis in section 4.3 suggests that a reduction of their power is not an immediate solution to it. To do so would certainly restrict excessive harmonization. However, it has been shown that the superior harmonization candidate is not always the one that is easier to reach.[133] Thus, when lowering harmonization power of these standards, excessive harmonization incentives may too often lead to harmonization with an inferior standard.

---

[131] See also Goerke and Holler (1998), p. 106: "As long as the commission cannot shift a greater share of the resulting costs of standard setting to producers or users, the European standardization organizations are motivated to produce more norms, in order to extend their bureaucratic institutions." See also Blankart and Knieps (1993) for similar derivation of incentives of standardization bureaucrats, underlined with historical examples.

[132] Such incentives seem to be well recognized. For example, clause 4.1.1 of the CEN/CENELEC internal regulations reads as follows: "CEN/CENELEC deal only with precise and limited subjects for which standardization is clearly and urgently needed [...]"

[133] See section 4.3.

Figure 20

a) Financing of the CEN

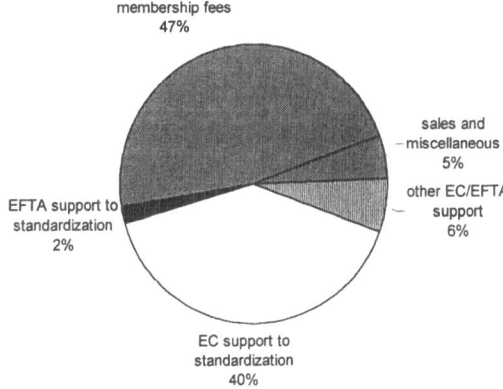

Source: CEN Annual Report (July 2000 to June 2001)

b) Financing of the DIN

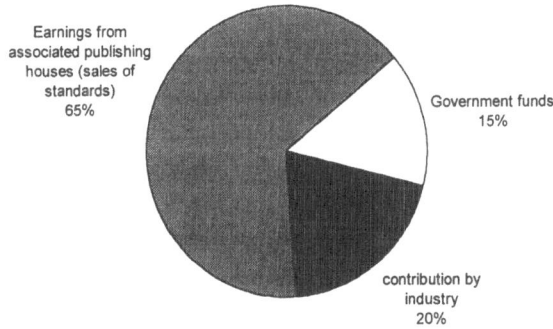

Source: DIN Annual Report 2000

4.5 Conclusions

Our analysis in section 3 suggests that globalization does not produce enough global standards.[134] Even if globalization does yield harmonization of national standards, there remain two pitfalls. First, harmonization typically occurs too late from a societal point of view. Second, populations may end up with an inferior global standard. In addition, we have found that the possibility of making double adoptions (adopting both standards) tends to support harmonization, even if they do not occur in (stable) equilibrium. Nevertheless, even though efficient harmonization is more likely to occur, it still comes with the risk that market-induced harmonization makes people end up with an inferior global standard.

There are several effects, potentially important ones, which our analytical framework has abstracted from. First of all, harmonization may have disadvantages that our welfare analysis does not account for. As pointed out by Blankart and Knieps (1993) among others, there is also a dynamic component involved. That is to say, harmonization might reduce innovation since it narrows the scope within market search for innovation is feasible. This might apply to "classical" goods as well as to institutions like contractual arrangements. Moreover, for deriving welfare implications, we have only compared the alternatives states. This might be insufficient, since we neglected costs that are certainly involved with transition from variety to harmonization, such as switching costs and possible lower welfare levels during the transition process. On the other hand, harmonization may have advantages that were not included in our framework. For example, harmonization may intensify competition, because it is easier for foreign firms to comply with national regulations and to compete with domestic firms.

In section 4, we have compared mandatory and voluntary formal standards as instruments to produce harmonization of global standards. From our evolutionary framework, we have derived a simple recommendation for harmonization policy: if at a particular harmonization project the "stability of variety" is low, a formal standard should be voluntary; if the stability of variety is large, a mandatory standard should be used.

We have found that EU policy for harmonization of technical standards makes, in fact, use of both kinds of instruments. Through granting an official and exclusive status to selected European standardization bodies, it fosters their power to generate collective switches through those standards, and, thus, it supports the use of voluntary formal standards. Moreover, in some cases, standards are directly or indirectly declared mandatory by European national legislations.

---

[134] The conclusions in this subsection belong to sections 3 and 4.

We have also found that this practice may lead to excessive harmonization of technical standards. This poses the question as to whether one should reduce the impact (the "collective switch power") of European official standardization bodies' standards. However, although this is likely to restrict excessive harmonization, it comes with the risk that these bodies would too often choose an inferior standard for harmonization.

5.   The Impact of Users' Commitments on Technological Progress in Network Industries

5.1   Introduction

5.1.1   Link to the Evolutionary Model in Section 3

This essay analyzes whether (inherently incompatible) new superior network technologies prevail over established ones, even though early switchers are worse off at the beginning. Although we consider one homogenous population, only, one should keep in mind that this problem can be similar to the harmonization problem: the analysis in section 3 has shown that agents might stick with their domestic standard, even if they are better off *if all of them* switched to the alien one. Obviously, in such a situation, harmonization is nothing more than transition to a superior standard.

One major source of variety's "excessive stickiness" in the evolutionary model is the assumption that agents behave myopically. In section 3.7.1 and 3.7.2, it has been argued that myopic behavior often coincides with ("rational") behavior under perfect foresight. Even under perfect foresight, players always choose the best response to the *current* distribution of standards if they can costlessly switch at any time and always know the present state. We have further argued that such myopic behavior might even be rational in cases where switching costs are involved. Even foresighted players choose the best response to the current distribution of standards if the adjustments process is slow and/or the interest rate is high. Such an adjustment process has not explicitly been modeled within the evolutionary framework in section 3, though. Instead, we simply assumed that agents tend to switch relatively fitter strategies over time, without taking account of the speed of such a process.

In this section, we change the assumption that players behave myopically and obtain an endogenous "adjustment process".[135] In such an environment, players anticipate dates of (potential) switches of other players and decide upon their own switch accordingly. Can a new superior network technology prevail over an established one under these assumptions?

Another major difference of the setting here to the former framework is that we explicitly consider the costs of transition from one state to another. Thus, even if a challenging technology is "statically" superior to an established one, it might be socially optimal if users stay

---

[135]   Of course, one could integrate an exogenous explicit dynamic process into the evolutionary framework. For example, one could assume that the distribution of strategies behaves according to a replicator dynamic, which also determines the rate of change for each date. See, *e.g.*, Holler and Illing (1996), p. 344ff, for details.

with the established one. Are established network technologies too sticky (excess inertia) or are they replaced too often (excess momentum)?

5.1.2 Summary of the Model

As a recent possible application of our specific setting consider the battle of the two competing TV-screen formats, the established 4:3 format versus the new superior 16:9.[136] Since human eyes tend to move easier sideways than up and down, the latter broader format is ergonomically better than the established narrower one.[137] It comes also closer to the format of cinema movies. Thus, it avoids black bars on the top and the bottom of the TV screen or parts of the movies simply being cut off. Due to imperfect compatibility of the two formats, (indirect) network effects are certainly involved here: the more people use a 16:9 format TV set, the more programs one can expect to be shown in that format (and vice versa).[138]

Due to imperfections of the market for used TV-sets, a buyer has to incur price drops if she wants to resell it. Moreover, there are costs of buying and installing a new TV-set. Thus, sunk costs are involved, making agents somewhat committed to the kind of TV standard once it has been chosen (see subsection 2.4). More precisely, people that still possess an old 4:3 format TV-set do not immediately buy a 16:9 TV-set as soon as the new format has passed a critical mass where its (net-) benefits have become larger than the benefits of the established one. The new TV technology's benefits have to exceed those of the established one by some extent, in order to compensate them for the costs associated with the "earlier" purchase of a new TV-set. On the other hand, such a commitment is often not perfect. Once sufficiently many programs are broadcasted in the 16:9 format (because many people possess a 16:9 TV) even those owning a still working 4:3 TV are willing to throw their old TV-set out of their living room in order to get a new 16:9 one. The relation of the network benefit functions of the two technologies and the users' levels of commitment determine when this threshold is reached. Intuitively, the weaker the users' commitments, the shorter is the transition time.

What makes transition actually *start*? Since a TV-set's lifespan is limited, a consumer, when his TV-set breaks down, is uncommitted from time to time. Transition only occurs if those

---

[136] Both formats are, of course, open standards.

[137] The larger a TV-set the stronger the effect. This is why cinemas screens and movies also possess a broad format. Given equal output, TV-set production costs do probably not differ between both formats. See also Meyer and Fontaine (2000).

[138] The two formats are only *imperfectly* compatible with each other, as one can still receive 16:9 programs even with a 4:3 screen – sacrificing some pleasure, of course. The problem is similar but still different to the introduction of color TV replacing black and white TV some 40 years ago. One crucial difference to the TV format case is that owners of black and white TV are not harmed at all, if TV stations send out programs for color TV, while, in our case, those possessing a 4:3 TV prefer programs to be broadcasted in 4:3.

users who become uncommitted are willing to be the first ones switching to the new technology, even though they are worse off at the beginning.[139] Imagine the problem of a consumer whose old TV-set has broken down just after the new technology has become available. Which of the two available formats will he choose?

The involved commitment is crucial for his decision, because it may render it costly to stick to the established technology. However, even if he expects that transition to the new technology will occur, he faces two alternatives. He can either switch to the new format now or stick to the established format and switch to the new one later[140]. The weaker the commitments are, the faster the transition will be. Thus weak commitments may support the alternative of adopting the new format now, because the costly period of incompatibility is shorter. On the other hand, the weaker the commitments, the easier (cheaper) to "jump on bandwagon", which makes the alternative to switch later more attractive. How do these effects trade off? As mentioned above, this decision is crucial for whether or not transition occurs in the first place. This implies that the decision of those agents have a major impact on social welfare. How does their optimal decision relate to the socially optimal one?

Our results for general technologies fitting our setting are summarized in Figure 21.[141] The horizontal axis measures the strength of agents' commitments, represented through C, defined as the costs of advancing the purchase of the new technology's asset (the 16:9 TV-set, for example).[142] The vertical axis measures the superiority of the new technology to the old technology.

The curve $W=0$ stands for superiority-commitment values where a transition to the new technology just improves social welfare. We find that transition to the new technology *actually occurs* if those agents whose assets brake down just after the introduction of the new technology are willing to switch to the new technology. Whether or not they are willing to switch, depends on two conditions. First, these agents' payoff stream must be larger when transition to the new technology occurs than if the population stayed with the inferior established technology – represented through condition PR^i, the "preference" condition. Second, these agents' payoffs are to be larger if they switch *now* instead of at any later date. This is accounted for in the "bandwagon condition", BW^i. Both conditions are just satisfied for values combined by the respective curves in Figure 21.

---

[139] Since in the model agents' preferences are assumed to be equal, one technology will eventually prevail.

[140] Such incentives are similar to those producing the "penguin effect" as introduced by Farrell and Saloner (1985).

[141] Figure 21 assumes linear "network benefit functions" (see section 5.3).

[142] See section 5.4 for details. In our specific setting, we assume that the net resale value of each good is zero. Thus, the users' commitments equal the costs of buying a new technology's asset earlier.

Figure 21

Agents' commitments, incentives to switch and desirability of transition

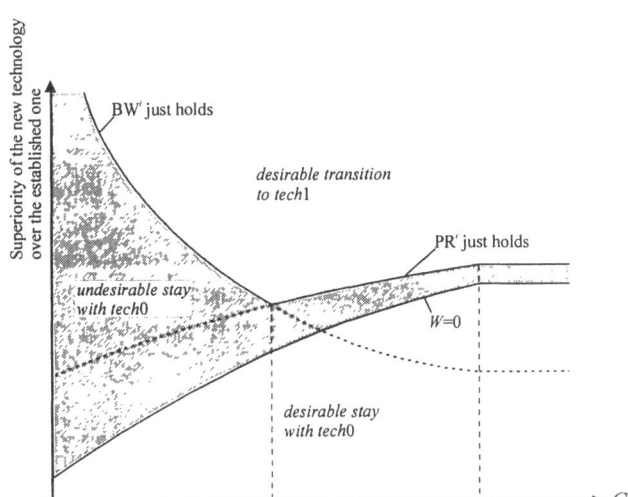

As indicated in Figure 21, superiority-commitment values located below $W=0$ imply that the population *desirably* stays with the established technology. Values above the BW^i *and* PR^i curves produce desirable transition to the challenging technology. Values that lie within the shaded area imply that the market fails to facilitate desirable transition, *i.e.* the population suffers from excess inertia.

C^H in Figure 21 stands for the threshold of C where it is so expensive for committed consumers to switch to the new technology that they only do so if their old technology's asset's life is (anyway) over. We find that for values beyond C^H, there is a slight tendency for excess inertia. Although, for general benefit functions, excess momentum is not impossible, linear network benefit functions imply that excess momentum cannot occur. We also find that for such large values of commitments, with linear network benefit function, transition always occurs if it is Pareto optimal.

For C-values between C^{crit} and C^H, as C becomes smaller, transition becomes both more likely and more desirable. Intuitively, the lower the commitments the faster the transition to the new technology will be. Thus, the costly initial period of incompatibility (the transition period) is shorter, which increases the transition benefits of every consumer. Nevertheless, since social transition benefits increase more than those of the first adopters, the risk of excess inertia is higher. If commitments fall below C^{crit}, the desirability of transition is even further augmented. However, for such low C-values, condition BW^i becomes binding. As it will be shown, when commitments decrease, an agents' incentives to pursue a jump-on-the-bandwagon strategy always increases, even though the transition time decreases. As a result, with respect to agents' commitments, transition tends to fail especially in those cases where it is highly desirable.

We further analyze the transition *time*. Even though, if below C^H, the transition process is shorter if commitments become lower, commitment values below C^H imply that the transition time is always too long. One reason is that agents who are still committed to the old technology are not able to coordinate a "premature" simultaneous switch to the new technology, even though such a collective action would make each consumer better off. If the transition process was accelerated, not only would social transition benefits increase but also the problem of excess inertia would be reduced.

Finally, it is asked whether policy-makers should intervene. We argue that, fortunately, there are appropriate means available especially when policy intervention is likely to be fruitful. It is proposed that policy-makers could introduce a small temporary subsidy in order to accelerate the transition process of new network technologies whose transition process *has already started*. With such a strategy policy-makers can circumvent the problems that are usually associated with policy intervention under the presence of network effects.

5.2   Review of Selected Literature

This essay is related to the literature on technological change in the presence of network effects. As described above, one focus of the study is how varying commitment levels of users relate to the private and social incentives of transition to new network technologies. Typically, literature contributions either assume that consumers' technology choices are perfectly irreversible or that consumers are not committed at all. Partly following thereof, the other focus of the study, the transition time in network markets, has also not widely been studied. While we ask whether the transition process is too fast or too slow, established papers ask instead whether the transition process *starts* too early or too late.

The seminal contribution on technological change in network industries is David (1985). As an example for excess inertia he considers key arrangements of keyboards. Although superior key arrangements are available, the inefficient QWERTY or QWERTZ keyboard remains standard. Various players (such as producers of keyboards, users and schools), the key arguments goes, are not able to coordinate a collective switch, and none of them is willing to make the first move. More theoretical contributions, in contrast, stress that network industries tend to suffer from excess momentum. One key assumption in these papers is that players are perfectly committed to their chosen technology.

5.2.1 Modeling dynamic demand

In order to differentiate our framework from established related models and to see which assumptions are crucial, we now discuss some theoretical contributions on the diffusion of new network technologies. We start with the demand side and come to supply and technology afterwards (section 5.2.2) where it is briefly discussed how things change if technologies are proprietary, *i.e.* exclusively marketed by one firm.

Table 4 structures how dynamic demand is modeled.

Table 4
Modeling dynamic demand

	OLG	Growing market	Fixed population	
			Two players	Large pop.
Exogenous sequence of adoption	Shy (1996) Fudenberg and Tirole (2000)	Regibeau and Rocket (1996) De Bijl and Goyal (1995) Arthur (1989) Farrell and Saloner (1986a) Katz and Shapiro (1986)	Farrell and Saloner (1986a)	Farrell and Saloner (1985)
Endogenous dates of adoption		Choi and Thum (1998) Koski (1998) Choi (1994) Katz and Shapiro (1992)	Moretto (2001)	This study Lange, McDade and Olivia (2001) Cabral (1990)

OLG and growing market

In a growing market model, new uncommitted buyers continually arrive and decide which technology to adopt: the established *tech0*, which typically already possesses an installed base, or the superior *tech*1, which has now become available. Other papers apply an overlapping generation model (OLG) where in each period, typically two generations of consumers coexist, an "old" one and a "new" one. At the beginning of each period, another generation of uncommitted buyers enters while the old one dies out and the previous new generation becomes the old one.

Both ways of modeling dynamic demand are similar to ours, as we also assume that agents are temporarily uncommitted.[143] However, most of these papers assume that agents have to make their decision (whether to adopt *tech*0 or *tech*1) at their time of arrival and, furthermore, that commitment is so strong that agents, once they have adopted a technology, are stuck forever with that which they have chosen. It seems intuitive that *tech*1 has a better chance of to becoming adopted than in our framework: due to perfect commitment, agents do not possess the option to jump on the bandwagon.

In fact, as Regibeau and Rocket (1996) and Katz and Shapiro (1992) show, excess inertia is rather exceptional in such a framework. To see why, note that excess inertia implies that it is socially better if newly arriving agents adopt *tech*1 but they choose *tech*0. However, if the social optimum requires that new agents adopt *tech*1, then it must surely yield more benefits to the new ones (making them adopt *tech*1) because their total benefits must even compensate for the foregone benefits of the installed base.[144] For the same reason it is possible (in some models even very likely) that excess momentum occurs.

In Choi and Thum (1998), Choi (1994), and Katz and Shapiro (1992), agents can postpone, however not advance, their technology adoption. In contrast, we assume that agents can advance and postpone their switch, but they *always* adopt *one* of the available technologies. In all three papers just mentioned, since adoption is perfectly irreversible, agents never first adopt *tech*0 and switch to *tech*1 once its network is large enough. Still, the option of waiting may either be in favor of the old or the new technology.

In the model of Katz and Shapiro (1992), waiting, however, never occurs in equilibrium. Consumers arriving when *tech*1 is already available, expect that either *tech*1 or *tech*0 will prevail. Consequently, there is no incentive to wait. Consumers that arrive before *tech*1 is available do not wait either. As Katz and Shapiro show, the firm offering *tech*0 has always an incentive to set prices low enough in order to capture all consumers until the firm offering

---

[143]   Which, in our framework, occurs when the life of the once adopted asset is over.

[144]   See De Bijl and Goyal (1995), pp. 308-309. The argumentation relies on the assumption that new arriving agents are able to coordinate on their "best" equilibrium, though. Thus, it is abstracted from coordination problems similar to the coordination games discussed in 2.2. See also sections 5.2.2 and 5.5.1.

*tech*1 enters the market.[145] In Choi and Thum (1998), building upon a model of Katz and Shapiro (1986)[146], the inferior *tech*0 is available before *tech*1 but does not possess a network ex ante. Consumers that arrive before *tech*1 is available must decide whether to adopt *tech*0 now (thereby enjoying a longer stream of benefits) or to wait and choose *tech*1 in the future. They find that consumers' incentive to wait tends to be too small. The reason is that "when the first generation of consumers adopts a technology too early, they inflict negative externalities on future consumers; either the future consumers are forced to buy an inferior technology or they lose compatibility in buying a superior technology" (Choi and Thum 1998, p. 226).[147] That is to say, due to the first mover advantage of the first generation and irreversible technology choice "forward externalities" tend to dominate "backward externalities". This result, however, clearly depends on their assumption that *tech*0 has no installed base. As demonstrated by Koski (1998), provided that *tech*0 already possesses perfectly committed users *ex ante*, the private incentives to wait may well exceed the social ones, *i.e.* backward externalities dominate (see Koski 1998, p. 60).

The OLG-framework of Shy (1996) – similarly to our framework – produces discontinuous adoption paths. In Shy's model, each new generation can adopt either a new incompatible technology or the established one, which has already been adopted by the old generation. According to some technology growth rate, for each *new* generation there is a better technology available. *Adopted* technologies' efficiency, in contrast, stays constant. A new generation of consumers is only willing to adopt a new technology if its inherently better quality compensates for the losses of network benefits (due to incompatibility of the new technology to the one adopted by the old generation). Consequently, from time to time, consumers refrain from adoption of a "superior" technology if the technology growth rate is not too high. Thus, discontinuous adoption paths arise. Shy, moreover, shows that the frequency of new technology adoption tends to increase if consumers treat the (stand alone) quality and network size as substitutes rather than as complements (see Shy 1996, p. 798).

Fudenberg and Tirole (2000) use a model similar to Shy's. Their focus, however, is on entry deterrence in network markets. Thus, they assume that technologies are proprietary, while Shy's (and our) model focuses on consumers' adoption choices given perfect competition on the supply side. (See section 5.2.2 for a brief discussion of how the evolution of standards changes if technologies are proprietary).

---

[145]   See Katz and Shapiro (1992), p. 63. The reason is that the date in which the *tech*1 firm enters the market is determined endogenously. The argument applies to competitive supply of each technology, too.

[146]   See subsection 5.2.2 for a discussion.

[147]   A similar intuition applies to Choi's (1994) framework.

Large fixed population

Let us now turn to markets where the number of consumers remains constant over time (fixed populations). The models of Farrell and Saloner (1986a, 1985) are related to ours.[148] In their 1985 paper, they investigate transition from $tech0$ to $tech1$ within a large population. They assume that all agents initially adopt the preestablished $tech0$. Agents, one after another, have the opportunity to switch to $tech1$. Transition to $tech1$ is the unique subgame perfect equilibrium. Applying backward induction, the player in the last period will switch if his predecessors have switched, and so the player in the period before the last one switches, which, in turn, is anticipated by *his* predecessor, etc. Although we apply a similar reasoning, there are major differences to our framework. First, Farrell and Saloner abstract from discounting. Thus, temporary incompatibility matters neither for private nor for social incentives.[149] Second, technology choice (at least adoption of $tech1$) is irreversible, which gives early switching agents strategic power through perfect commitment. Third, since switching always occurs sequentially, their model does not allow for an analysis of the role of transition time, which has a great deal of impact within our framework, as dates of switches are determined endogenously.

In Cabral (1990), endogenous dates of adoption may produce a discontinuous adoption path. Cabral, though, abstracts from installed base effects. In essence, additional to network effects, two assumptions generate an adoption path as displayed in Figure 22. First, net benefits of $tech1$, for *given* number of users, increase with time. Second, consumers' preferences are heterogeneous.

The horizontal axis in Figure 22 measures the time; the vertical axis gives the share of users of $tech1$, denoted by x. For low values of t, only those agents with a high valuation of $tech1$ choose to adopt. For intermediate t-values, multiple equilibrium shares exist; the reasons have already been discussed in section 2.2. The downward sloping part of the curve corresponds to equilibria in mixed strategies similar to those in the pure coordination game displayed in Figure 2. This part slopes downward because $tech1$'s efficiency increases with t. Thus, in order to keep agents indifferent the probability (or share) put on adoption of $tech1$ must be smaller when t increases. Once t has passed some threshold (t'), adoption of $tech1$ by most (or all) agents is the only "static" equilibrium.[150]

---

[148]   In part I of Farrell and Saloner (1986a), a model with a growing market as discussed above is analyzed.

[149]   Together with the exogenous assumption of sequential moves, this always makes transition the unique subgame perfect equilibrium if the new technology is superior.

[150]   Less sophisticated benefit functions produce equilibrium shares where for low t-values no agent adopts $tech1$ and for high values any agent adopts $tech1$.

Figure 22

Equilibrium adoption path in Cabral

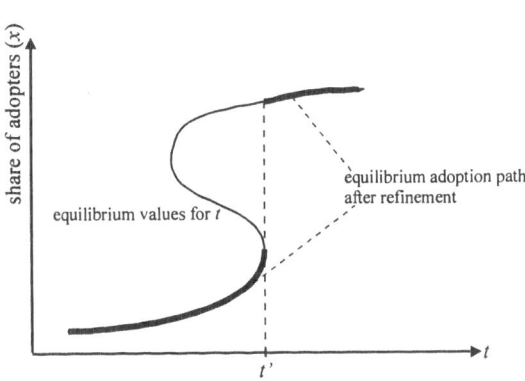

Source: Cabral (1990), p. 303.

To select among x-levels at intermediate t-values, Cabral assumes that agents' adoption decision at any t is based on their observation of x at t-ε, where ε is arbitrarily small (see Cabral 1990, p. 302). That is to say, agents expect that the share of adopters at t is very close to the one observed just before adoption. Consequently, the adoption path is given by the lower envelope of the curve (the thick line) in Figure 22, and so the adoption path involves a discontinuity at t'. This result of Cabral is related to one of our finding, as our framework may also produce discontinuous adoption paths. Although we use a somewhat similar way to select among equilibria, Cabral's method involves the problem that agents' expectations are not satisfied at t': the equilibrium share at t' substantially differs from the share at t-ε.

Lange, McDade and Olivia (2001) claim that it is difficult to model discontinuous adoption paths when players behave according to "smooth" decision rules.[151] (Cabral's framework as well as Shy's 1996 and ours give a counterexample.) Instead, they use a "catastrophe model", which, in essence, turns out to be an extension of the above model of Cabral, allowing for varying degrees of network effects' strength.

A substantive difference of our framework to both Lange, McDade and Olivia's and Cabral's is that in those models agents are never worse off in the first periods after their switch. Agents adopt the emerging technology only when it immediately provides for larger payoffs (than an outside option).

---

[151] See Lange, McDade and Olivia (2001), p. 31-33.

Models with two players

In their 1986a paper (part II), Farrell and Saloner include discounting into their 1985 model. However, they also reduce the population of users to two players. Again, agents can only switch if afforded the opportunity, which arises exogenously (now stochastically) from time to time. Farrell and Saloner demonstrate that in such a setup, private incentives to switch may or may not produce efficient transition. For some parameterizations transition occurs and is efficient. For other parameter values excess inertia occurs; either in·a strong form where both agents stick with *tech*0 although transition is efficient, or in a weak form where transition occurs too late. The reasons for excess inertia in the strong form are due to the "penguin effect" as already discussed. The delay in the latter case occurs if agents adopt mixed strategies.

Agents may also suffer from excess momentum, *i.e.* both agents switch to *tech*1 although social welfare would be maximized if they stayed. Excess momentum may occur because an agent who contemplates switching takes into account only his own losses from temporary incompatibility. For efficiency, he should ask for losses of both agents (see Farrell and Saloner 1986, p. 954).

Most of their results apply in our framework, too. However, our focus is on the analyses of the effects of adopters' commitments to the chosen technologies, which Farrell and Saloner do not consider. Second, due to the fact that the dates of switches are derived endogenously, our model also allows for the analysis of transition time. Third, due to the two-player-setup, transition to *tech*1 is easier than in our multiple user setup, because the first switcher sacrifices relatively fewer initial network benefits than in the case where the population is large.

Moretto (2001) also models a two player setting similar to Farrell and Saloner (1986a). As in our model, players' dates of switch are derived endogenously. Since, in their model, switching is costlier for the first switcher, each of the two players prefers to be the second switcher.[152] On the other hand, it is assumed that *tech*1's efficiency increases over time. This, together with the assumption that the opponent's switching costs are unknown, generates a "war of attrition" game in which both players tend to switch too late compared to a cooperative setting.[153]

---

[152] To be (perhaps too) precise, although this exogenous assumption is used to capture network effects, this assumption does not go along with the definition of network effects as used in our study. In fact, the benefits with each technology in Moretto (2001) are independent of the number of users. In part 4 of that paper, benefits even decrease with network's size. In this setting, a preemption effects counteracts delay of transition.

[153] See Fudenberg and Tirole (1991), chapter 4.5.2, Peters (1998), chapter 2.2.3, for details on the war of attrition game.

5.2.2 A Brief Note on Sponsored vs. Non-Sponsored Technologies

Let us now turn to supply and technology. One crucial assumption in our model is that both technologies are "non-proprietary" or "non-sponsored". A technology is said to be sponsored if a firm exclusively markets it. Note for clarification, what matters here is that other firms *cannot* offer *compatible* technologies. Either other firms are simply not able to do so (*e.g.* for technological restrictions) or the firm possesses sufficiently broad and enforceable intellectual property rights for its technology.

Liebowitz and Margolis (1994) argue that excess inertia should be more the exception. "A transition to a standard or technology that offers benefits greater than costs constitutes a profit opportunity for entrepreneurial activities that can arrange the transition and appropriate some of the benefits" (Liebowitz and Margolis 1994, p. 146). It is clear that it is costly for a firm to facilitate such a transition process, since in order to do so it must make introductory offers, advertising expenses or undertake similar measures. Thus, making use of such profit opportunities requires that the technology is sponsored. Otherwise, firms that can offer *compatible* technologies would have an incentive to free-ride on other firm's transition expenses.

The seminal and often employed paper of Katz and Shapiro (1986) can provide a theoretical basis for the argument of Liebowitz and Margolis. This is the setting. There are two periods, t_0 and t_1. In t_0, the first cohort of consumers enters the market followed by the second cohort in t_1. In t_0, *tech0* is superior to *tech1*, and conversely in t_1, *tech1* is better than *tech0*. Consumers' technology choice is perfectly irreversible.

If technologies are not sponsored, *tech0* has a first mover advantage. Since *tech0* is superior in t_0, t_0-consumers have stronger preferences for *tech0* being the standard, than t_1-consumers. They choose first and lock themselves into their preferred technology and influence t_1-consumers in choosing *tech0* as well. In contrast, if both technologies are sponsored, then there is a second mover advantage, since the sponsor of *tech0* cannot credibly commit to set sufficiently low prices in t_1.

To consider these effects in more detail, let a_{it}, $i,t = (0,1)$, denote the stand-alone value derived from adoption of *tech i* in period t. If both cohorts adopt the same technology, consumers receive an additional benefit of n.[154] Denote Δ_t the advantage of either superior technology at t_0 and t_1, respectively: $\Delta_0 = a_{00}-a_{10}$ and $\Delta_1 = a_{11}-a_{01}$. Suppose that $\Delta \equiv \Delta_0 = \Delta_1$, implying that utilitarian welfare is equally large regardless of whether all consumers adopt *tech0* or *tech1*.[155]

---

[154] The results remain unchanged if t_0-consumers' benefits take the form $a_{i0} + a_{i1} + n$; see Katz and Shapiro (1986, p. 826).

[155] Total costs are zero for both technologies. Note that the benefit functions imply that it is abstracted from discounting.

However, t_0-consumers prefer harmonization with *tech0* and t_1-consumers prefer harmonization with *tech1*. Further suppose that n is sufficiently large such that t_1-consumers always adopt the technology chosen by the first cohort.

The fact that there is a first mover advantage of *tech0* in the case of non-sponsored technologies is obvious and so are possible inefficiencies. The second mover advantage of *tech1* in the case of sponsored technologies can be derived with backward induction. Profits in t_1 of price-competing sponsor 0 and sponsor 1 (π_{11}, π_{01}), respectively, *given* that they have captured all t_0-consumers, are:

(E16a) $$\pi_{11} = a_{11} + n - a_{01} = \Delta + n$$

(E16b) $$\pi_{01} = a_{01} + n - a_{11} = -\Delta + n$$

The profit opportunities in t_1 make the sponsors fiercely compete for the t_0-consumers; in order to get these consumers, both sponsors are willing to sacrifice their entire potential t_1-profits. Sponsor 1's "budget" exceeds sponsor 0's by 2Δ. However, to attract t_0-consumers he only needs Δ more than sponsor 0. Thus, *tech1* possesses a second mover advantage. This also implies that *tech1* may inefficiently prevail (*e.g.* if Δ_0 is just larger than Δ_1), *i.e.* there is a tendency for excess momentum.

Sponsor 0 suffers from t_0 (where his technology is better) being the "competitive" period whereas sponsor 1's period is the "milking" one. Things would change if sponsor 0 could commit to low prices in t_1. If sponsor 0 could guarantee to t_0-consumers that he will later attract consumers in t_1, any asymmetry produced by the period's sequence would diminish, and so would any first or second mover advantage.[156]

If only *tech1* is sponsored, things change because competition drives the prices of *tech0* in both periods to 0. Now sponsor 1 has a problem to commit to low prices in t_1.[157] But still, if he has an efficiency advantage over *tech0* he can, in extreme cases, sacrifice his entire t_1-profits in t_0 in order to make his technology succeed. This strategic option, which *tech0* does not have without a sponsor, does not only compensate for his inability to commit to future prices; even if *tech0* is efficient, it is possible that *tech1* prevails in both periods. Responsible is the "weakened rival effect": by winning in t_0 sponsor 1 faces a rival that is less attractive

---

[156] See Thum (1995) and Katz and Shapiro (1994) for techniques available to sponsor 0 to commit to low prices.

[157] In the above-discussed model of Choi and Thum (1998), the sponsor's inability to commit to future prices produces the further disadvantage that incentives for efficient waiting of t_0-consumers is further reduced.

for t_1-consumers, and thus sponsor 1 can appropriate more of the potential surplus offered to them. This transfer is a private, but not a social, benefit of sponsor 1's winning the first period (see Katz and Shapiro 1986, p. 835).

As a formal note, notice that the outcomes of the Katz and Shapiro model depend on the assumption that agents within a period are able to coordinate on "their best" equilibrium.[158] That is to say, coordination problems such as given in the "simple coordination game" (see Figure 2 in subsection 2.2) are assumed to be solved. Consider the game without sponsoring, for example. Without Katz and Shapiro's assumption, harmonization with $tech1$ is a subgame perfect equilibrium, too. To obtain this path, it must only be assumed that t_0-consumers expect that all other t_0-consumers choose $tech1$ in t_0. Subgame perfection would not rule out such expectations, as choices within t_0 are to be made simultaneously. Nevertheless, in our model, a similar assumption is used.

The model of Katz and Shapiro shows that the presence of sponsors of technologies has a great deal of impact within network markets. Sponsors can internalize some of the externalities generated through network effects. Still, internalization often remains imperfect and strategic options of a sponsor may even produce other inefficiencies. Katz and Shapiro's model suggests that sponsoring tends to produce more excess momentum in network markets. In general, it remains however unclear whether sponsored or non-sponsored network markets perform better.

Why a model with non-sponsored technologies?

In our study we choose a framework where all (both) technologies are non-sponsored. There are several reasons why this case is relevant.

• As mentioned in the beginning of this section, we can use the framework to ask whether "the market" induces harmonization when integration has turned an alien standard superior to the domestic one. Thus, the "superior" standard or technology can very well have been established in the alien population for a long time. Consequently, patents or copyrights as well as some innovator's first mover advantage (*e.g.* due to initial superior specific technological expertise) may already have expired.

• There are many network goods that are not allocated through typical markets. What we have in mind are "goods" like contract-terms, social norms, languages etc.

• Many scholars such as Lemley and McGowan (1998a) and Farrell (1995) plead for weaker intellectual property rights for network technologies; especially, interface protec-

---

[158] This assumption has become a standard one in network effects literature. See, *e.g.*, Fudenberg and Tirole (2000), p. 377.

tion should be denied. It is, however, exactly the interface protection that keeps other firms from offering compatible technologies. Thus, as argued above, given innovator's technical first mover advantage is not sufficient, firms' incentives to incur expenses that facilitate transition is low. In fact, there are many network goods whose interface is not protected by intellectual property law. For example in *Lotus Dev. Corp. v. Borland Int'l*, the US First Circuit finds that the command hierarchy in Lotus's "1-2-3"-spreadsheet is not copyrightable (see also Lemley and McGowan 1998a, p. 81). As another example, consider the two TV screen formats from the introduction.

5.3   The Model Setup

Assume that there is a large population of users ("agents"), approximated with a continuum of size 1. Currently, all agents apply the same technology, namely *tech0*. Now, at date $t=0$, a new, superior technology (*tech1*) has evolved and become available to every agent. We are interested in whether and under which conditions there will be a transition to *tech1* and, if so, how long the transition takes.

The (gross) benefits an agent enjoys at date t depend on two variables. The first one is the technology he applies and the second one is the number of users of that technology. Precisely, benefits are given by $a_i^t(x^t)$. Thus, the benefits of an agent at the date t using *tech i*, $i = (0,1)$, depend on x^t, the share of users of *tech1* at date t, $t \in [0,\infty)$; the remaining agents, $1-x^t$, use *tech0*. Network effects imply that $\delta a_1^t(x^t)/\delta x^t \geq 0$ and $\delta a_0^t(x^t)/\delta x^t \leq 0$. It is assumed that $a_i^t(x^t) \geq 0$, $\forall x^t \in [0,1]$. This also entails that both technologies exhibit non-negative "stand-alone benefits", which are denoted by a_0^0 and a_1^0, respectively. Furthermore, it is assumed that the application of technologies is inherently mutually exclusive (see sections 2.3, 2.5 and 3.8 for discussions).[159]

There are also costs coming with the application of a technology. I_i denotes the investment that is required for application of *tech i*. We can say that a user of each technology has to acquire a durable good (referred to as an "asset") that enables him to apply that technology, such as a TV-set to enjoy watching TV. I_i also includes transaction costs that are associated with the investment, such as costs of assessing, ordering and installing the new TV-set and getting rid of the old one.

---

[159]   It says that agents never adopt *both* technologies at the same time. Using the above example, we assume that people never put two TV-sets, one having the 4:3 and another one with the 16:9 format, into their living room. This should apply to most people as they probably evaluate a nice living room higher than perfect TV compatibility.

For simplicity, we assume that these costs, together with a discount on used assets, ensure that the entire investment is sunk (see section 2.4 for discussion.).[160] We only consider technologies whose net benefits are larger than an outside option. Together with the assumption that technologies are inherently mutually exclusive, this insures that *every* agent *always* adopts *one* technology.

Figure 23

Distribution of investment schedules and transition benefit functions

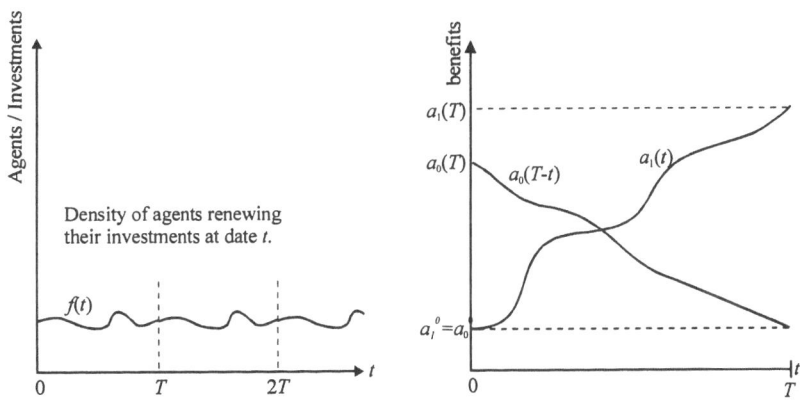

After some time has passed, assets need to be replaced, *i.e.* another I_i is due. To keep it simple we assume that an asset of *tech i* has a lifespan of T_i and $T := T_0 = T_1$. So each agent, irrespective of the technology he applies, has to make a new investment at least once in a time period with a length of T (referred to as a "round").

Although agents are well informed and the assets' lifespan is not stochastic, it remains a natural assumption that, at the starting point, the age of the currently adopted asset differs among agents. There are several possible reasons. For example, individuals were born on different dates or, if agents stand for companies, they were founded at different dates. Depending on the specific technology, mergers and breakups (marriages and divorces, respectively) should further lead to smooth distributions of the assets' age. One may conclude that the longer *tech0* is in place the smoother the distribution of the assets' age.

---

[160] The assumption that there are sunk costs involved is crucial for the analysis. However, that the *entire* investment is sunk is only a simplification, the qualitative results of the model hold true even for partially sunk investments. Note, if there are costs of reselling or otherwise getting rid of the old asset it may still have some positive market value.

In Figure 23 *(lhs)*, $f(t)$ gives the density of agents that have to renew their asset at $t=[0,T]$. On average, $f(t) = 1/T$. Since the assets' life is non-stochastic, $f(t)$ is identical in each round. $f(t)$ together with $a_i'(x)$ produce the curves $a_1(t)$ and $a_0(T-t)$, respectively, which are referred to as "transition benefit functions". These are shown in Figure 23 *(rhs)*. The curves give the benefits enjoyed at t by agents who apply *tech i*, $i=(0,1)$, provided each agent switches from *tech0* to *tech1* at the date where his *tech0*'s asset's life expires. As a consequence of network effects, $a_1(t)$ increases monotonically and $a_0(T-t)$ decreases monotonically. Otherwise both curves' shapes are arbitrary. In the figure, both technologies' stand-alone benefits, a_0^0 and a_1^0, are equal.[161]

As an example of a specification of the transition benefit functions, assume that reinvestment dates are equally distributed, *i.e.* $f(t) = 1/T$, $t=[0,T]$, and assume linear network benefit functions of the form $a_0(1-x) = a_0^0 + (a_0-a_0x)$ and $a_1(x) = a_1^0 + a_1x$, respectively. These produce transition benefit functions $a_0(T-t)$ and $a_1(t)$, respectively:

(E17a) $a_0(T-t) = a_0^0 + a_0-(a_0/T)t \quad, t=[0,T]$

(E17b) $a_1(t) = a_1^0 + (a_1/T)t \quad, t=[0,T]$

We now make the two assumptions that are responsible for the existence of the problem in our framework. The emerging *tech1* is superior to the established *tech0*. That is to say, with (any) *equal* number of users, the net benefits exhibited by *tech1* are greater than those of the established *tech0*. However, if the number of users of *tech0* is larger than the number of users of tech1, the net benefits of *tech0* *might* be larger than the net benefits of *tech1*. As a minimum requirement we state that *tech0*'s net benefits are always larger than *tech1*'s if every users adopts *tech0*.

Assumption (5.A) $-I_1 + \int_0^T a_1(x'')e^{-rt}\,dt \geq -I_0 + \int_0^T a_0(x'')e^{-rt}\,dt \quad \forall x' \in [0,1], t$

Assumption (5.B) $-I_1 + \int_0^T a_1(0)e^{-rt}\,dt < -I_0 + \int_0^T a_0(1)e^{-rt}\,dt$

---

[161] While most of the figures below assume this, it is not included in the analytical analysis except for the welfare analysis in section 5.5.

where r is the common interest rate. Assumption (5.B) produces the potential collective action problem. Obviously, if $tech1$'s (net) stand-alone value exceeded $tech0$'s full network value, transition always would start as soon as $tech1$ is available. Other assumptions pertaining to costs and benefits are not made. So in relevant variables, as $e.g.$ the stand-alone benefits (a_0^0 and a_1^0), $tech0$ may very well exhibit more benefits than $tech1$.

Assets of both technologies are competitively supplied. Hence, we abstract from strategic interactions between suppliers of the assets pertaining to both technologies. This assumption rules out suppliers' measures such as sponsoring of a technology as discussed above.

We assume that both the time and the number of agents are continuous variables. One can think of the continuous time being the limit of a sequence of very short periods where at the beginning of each period, all agents decide simultaneously about the technology to adopt.[162] Continuous time is certainly a reasonable assumption, as it allows agents to buy each kind of asset at any time. A continuum of agents just approximates a large population of agents such that actions of one single agent do not relevantly affect the payoffs of other agents.

There are a lot of "industries" that fit these assumptions. One might think of typical consumer home entertainment technologies such as DVD vs. VCR, Mini Disc vs. CD, or (as discussed in the introduction) TV format 16:9 vs. the established 4:3.[163] The model may also be applied to production technologies where network effects might be caused by availability of service, expertise and staff. In a broader interpretation the framework may also help to analyze the diffusion of social and legal norms and contract terms if they are subject to network effects. Of course, such "goods" do not suffer from depreciation. However, opportunities for easy switch may arise from time to time. For example, a public stock company may find it easier to change its corporate charter terms if it nevertheless holds a periodical shareholder meeting.

The remainder of this essay is organized as follows. Section 5.4 deals with the equilibria of the game. Section 5.5 asks how the market performs, $i.e.$ whether and how much the game's outcome differs from the socially optimal one. Section 5.6 briefly discusses policy implications. Finally, section 5.7 concludes.

5.4 Equilibria

It may make sense to start this section with a brief note on its structure, as it might – at first glance – appear quite peculiar. The first subsection 5.4.1 deals with the existence of three selected equilibria: the entire population "perpetually" stays with the established technology

---

[162] See Fudenberg and Tirole (1991), p. 520.

[163] In the previous two examples, there has probably been some (indirect) sponsoring involved, though.

(NE^{est}), the entire population switches immediately to *tech*1 (NE^{rap}), the population switches "sequentially" to the challenging technology (*i.e. each* agent switches at the date where his old asset breaks down), denoted by NE^{seq}. In deriving the conditions for the existence of these equilibria, we develop our modeling of the agents' commitments.

As a next step we complete our equilibrium analysis, which is divided according to the strength of the agents' commitments. In section 5.4.2, the commitments are "strong".[164] It is shown that strong commitments imply that no other equilibrium (path) than NE^{est} and NE^{seq} exists, and we demonstrate that for each parameter constellation, there is a unique "plausible" equilibrium. In sections 5.4.3, we deal with the residual cases (low commitments). We find that a continuum of equilibria exists. Nevertheless, we argue that for each parameterization, a unique plausible equilibrium exists. Typically, this equilibrium starts with sequential adoption followed by simultaneous switch by the rest of the population at some "interior" date (which depends on the specific parameter values).

5.4.1 Perpetual Stay, Immediate Switch and Sequential Adoption

In this section, we check whether and when the following three adoption paths exist as Nash-equilibria:

A) The population stays with the established technology perpetually (NE^{est}).

B) The entire population switches immediately to the new technology (NE^{rap}).

C) The agents switch sequentially to the new technology (NE^{seq}).

At each date *t*, agents simultaneously decide which technology to apply. At *t*=0, the new technology (*tech*1) has become available. So, agents decide whether to stick with the old *tech*0 or to invest in the new *tech*1. More precisely, they decide whether to switch now or later. However, except for "extreme" shapes of $f(t)$, the agents' situations differ with regards to the date at which they have to renew their asset. The agent who needs to replace his asset at date *i* is labeled agent *i*. For example, agent 0 has to make a reinvestment at *t*=0 and agent *T* at *t*=*T*.

Whether the decisions "later" and "never" are conditional choices, *i.e.* conditional on observations regarding the previous choices of the other agents, depends on the information structure. Since the benefits depend on the number of agents applying the same technology, our benefit functions suggest an information structure according to the concept of "closed-loop",

---

[164]   There is a threshold beyond (below) which commitments are denoted "strong" ("low").

i.e. each agent knows the present state at each date. Thus, strategies entail conditioning of actions at each date on the history of that date.[165]

A) Perpetual stay (NE^{est})

If agent 0 expects that *no* (other) agent will *ever* switch to *tech*1 he will stay with *tech*0 if

(C8a)
$$-I_1 + \int_0^T a_1(0)e^{-rt}\, dt < -I_0 + \int_0^T a_0(T)e^{-rt}\, dt$$

Computations, substitutions and rearrangements yield

(C8b)
$$a_0(T) - a_1(0) > (I_0 - I_1)r\frac{1}{1-e^{-rT}}$$

When we define

(DIV)
$$C_i := I_i r\frac{1}{1-e^{-rT}}$$

condition (C8b) becomes

(C8c)
$$a_0(T) - a_1(0) > C_0 - C_1 \qquad {}^{166}$$

Condition (C8a) is equivalent to Assumption (5.B). It states that agent 0's first round's compounded net benefits when solely using *tech*1 must be smaller than his net benefits when he applies *tech*0 together with everyone in the population. (The integrals accumulate the present values of the benefits of *tech*0 with a full network and the stand-alone benefits of *tech*1, respectively.)

We need to consider only the first round because at the beginning of the next round (right after *t=T*), agent 0 has again the opportunity to switch to *tech*1 without incurring additional costs. If condition (C8)[167] holds for the first round, an equivalent condition will hold for each subsequent round.

---

[165] We confine our attention to pure strategies.

[166] Note that this condition turns into $a_0(T) > a_1(0)$ if $I_0 = I_1$.

[167] If we refer to a "group" of equivalent conditions we omit the specific letters.

Since every other agent is even relatively worse off through adopting *tech*1 at $t=0$, no agent switches at $t=0$ if (C8) holds. Afterwards, at $t=0+\varepsilon$, the "next" agent's investment becomes due. Since he faces exactly the situation which agent 0 has faced at $t=0$, he also refrains from adoption of *tech*1 and stays with *tech*0 (if agent 0 has done so). The same is also true for the "next after the next" agent, etc.

To conclude, as long as Assumption (5.B) is given, it is a Nash-equilibrium that no agent ever switches to the new technology (NE^{est}).

Note that the *rhs* of condition (C8b) represents the difference of the (average) costs of capital of the technologies per "time-unit". C_i, as defined in (DIV), denotes the "average cost of capital" (including depreciation) associated with applying technology i. These are the "equivalent annuity" (in a continuous formulation) of I_i with respect to a time period with a length of T. The parameter C_i has an intuitive interpretation: it equals the asset's constant leasing rate that agents would have to pay "at each t", given competitive financial markets.

B) Immediate simultaneous switch (NE^{rap})

Now, let agents expect that all (other) agents immediately switch at $t=0$, *i.e.* right after *tech*1 has been introduced. Since agent 0 has to make a reinvestment anyway and *tech*1 is superior to *tech*0, he switches to *tech*1 at $t=0$. To establish this "rapid" diffusion as a Nash-equilibrium, every agent (expecting that all agents do so) must switch. Agent i switches at date $t=0$ rather than at date $t=i$ if

(C9) $$-I_1 \frac{1-e^{-ri}}{1-e^{-rT}} + \int_0^T a_1(T)e^{-rt}\,dt > \int_0^i a_0(0)e^{-rt}\,dt + \int_i^T a_1(T)e^{-rt}\,dt$$

In condition (C9), the second term of the *lhs* gives the present value of *tech*1's full network benefits for the first round. Since everyone switches to *tech*1 at $t=0$, the benefits stay constant throughout the entire round. On the *rhs* of (C9), the benefits of the best alternative are shown. Clearly, if everyone else adopts the new technology immediately in $t=0$, agent i's best alternative is to follow at $t=i$. To follow at a date located between 0 and i can never be optimal, as costs and benefits are constant, so either the benefits are higher (switch at $t=0$) or lower (switch at $t=i$) than the costs. (Of course, to switch later than at $t=i$ cannot be optimal because this implies another reinvestment into the inferior tech0 at $t=i$.)

Agents' commitments

The first term in (C9) stands for the additional costs of capital that agent i has to incur if he switches at $t=0$ instead of at $t=i$. With switching at $t=0$ instead of at $t=i$, agent i throws away his still working asset, *i.e.* he sacrifices the value of possessing an asset which would otherwise have worked for another period of i time-units.[168] This term also includes the effects on all subsequent rounds. That is, if agent i invests i time-units earlier in the first round, he will have to replace his assets i time-units earlier also in each subsequent round.

Differentiating this term with respect to i yields

(E18)
$$\frac{\partial\left(-I_1\,\frac{1-e^{-ri}}{1-e^{-rT}}\right)}{\partial i} = C_1 e^{-ri}$$

These are the compounded costs of making an investment "one time-unit" earlier. Situated in $t=i$, these amount to C_1, the "average costs of capital as derived above. Since it is equally costly for each agent at any given time to make an investment one time-unit earlier, C_1 also represents the *"marginal"* costs of capital. That average and marginal costs of capital are equal is due to the infinite time horizon that agents face. Remember, in another interpretation, C_1 equals the constant leasing rate to be paid for a *tech*1-asset. If agent i switches at $t=0$ instead of at $t=i$, during $t=[0,i]$ he has to pay leasing rates for *both* *tech*0's and *tech*1's asset.[169] (Of course, since the costs for the assets are entirely sunk, an agent cannot save money through giving back the old asset to the leasing firm and cancel the leasing contract.)

Since C_1 stands for the costs of advancing the switch to the new technology, it represents the agents' level of commitment to the once adopted *tech*0. In other words, C_1 represents the extent to which the benefits of *tech*1 exceed the benefits of *tech*0, whereby causing an agent to discard his still working *tech*0 asset and procure a new *tech*1 asset. Thus, in our framework, C_1 measures the agents' level of commitment. Of course, if the assets were only *partly* sunk, the levels of commitment would reduce according to the assets' (net) resale value.

To check whether NE^{rap} exists, it suffices to consider agent T's choice (since costs and benefits remain constant).[170] For him, condition (C9) is identical to condition (C10a). That is, agent T switches to *tech*1 at $t=0$ iff

---

[168] Remember that we assume that the investment is completely sunk.

[169] Remember that the assets' net liquidation value is 0; otherwise C_1 (the leasing rate) must be reduced by an amount corresponding to the resale's net returns.

[170] We exclude the possibility of waiting for the new technology through sacrificing benefits of a period. That is, although *tech*1 evolves right afterwards, agent T just invested into technology 0. So, we implicitly assume that the net benefits are large such that it never pays off to wait. Alternatively, we could also assume that it had been sufficiently uncertain when the new technology becomes available. Effects generated by agents' option to wait have briefly been discussed in section 5.2.1.

(C10a)
$$-I_1 + \int_0^T a_1(T)e^{-rt}\, dt > \int_0^T a_0(0)e^{-rt}\, dt$$

which is equivalent to

(C10b)
$$\frac{1-e^{-rT}}{r}[a_1(T) - a_0(0)] > I_1$$

Plugging in (DIV), this becomes

(C10c)
$$a_1(T) - a_0(0) > C_1$$

Condition (C10a) states that agent T switches, if the value of the first round $tech1$ benefits with a full network minus an additional investment exceeds the value of the first round stand-alone $tech0$ benefits. In condition (C10b), the factor before the brackets accumulates and discounts agent T's additional benefits; his additional costs are an entire investment at $t=0$, i.e. I_1. This condition is equivalent to (C10c) which says that the additional benefits per "time-unit" (i.e. $tech1$'s full network benefits minus the $tech0$ stand-alone benefits) have to compensate for the agents' commitments, being equivalent to $tech1$'s (marginal) costs of capital.

Since condition (C10c) applies to each agent, we can conclude: if condition (C10c) holds, then an immediate simultaneous switch of the entire population is a Nash equilibrium (NE^{rap}).

C) Sequential Adoption

Now we want to know whether an equilibrium exists that entails the *entire* population switching to the new technology "one after another", i.e. sequentially (NE^{seq}).[171] Precisely, each agent switches only when his investment is due, i.e. agent i switches from $tech0$ to $tech1$ at $t=i$. No agent incurs additional costs of capital since he has to replace his asset at date i anyway. In order to establish this pattern as an equilibrium, no agent may benefit from making his investment earlier or later. We first look for conditions, which insure that no agent wants to *advance* his switch. Afterwards the condition is established under which neither agent has an incentive to *postpone* his switch.

With sequential adoption, the net benefits of agent i are

---

[171]   As far as the term sequential applies to continuous set of agents and time.

$$(E19) \quad u^{seq}(i) = \int_0^i a_0(T-t)e^{-rt}\, dt + \int_i^T a_1(t)e^{-rt}\, dt - I_1 e^{-ri} + \frac{a_1(T)}{r}e^{-rT} - \frac{I_1 e^{-r(T+i)}}{1-e^{-rT}}$$

From $t=0$ to $t=i$ agent i receives benefits according to $a_0(T-t)$. At $t=i$, he switches to *tech*1. From then until $t=T$ he gets $a_1(t)$. From date T onwards he gets constant benefits; the value of that stream is given by the last but one term in condition (E19). $I_1 e^{-ri}$ stands for the present value of the first round investment at $t=i$, and the last term sums up the values of the investments of all subsequent rounds.

Agent i switches only (not earlier) at $t=i$ iff $\delta u^{seq}(i) / \delta i \geq 0$. If $\delta u^{seq}(i) / \delta i < 0$, agent i would be worse off than an agent located before him. Then, however, sequential adoption could never be an equilibrium since agent i could simply make his investment earlier. If $\delta u^{seq}(i) / \delta i \geq 0$, agent i is better or equally well off than any of his predecessors, and so he will invest not earlier than at date i. Thus, a *necessary* condition for the sequential adoption schedule to be a Nash equilibrium is

$$(C11a) \qquad \frac{\partial u_i^{seq}}{\partial i} \geq 0 \quad \Leftrightarrow \quad I_1 r \frac{1}{1-e^{-rT}} \geq a_1(i) - a_0(T-i) \quad , \forall i$$

or, after including (DIV):

$$(C11b) \qquad C_1 \geq a_1(i) - a_0(T-i) \quad , \forall i$$

Conditions (C11) have an intuitive interpretation. If agent i makes his investment one "time-unit" earlier, he obtains three effects. First, he imposes additional capital costs on himself, since he has to finance the assets of *tech*1 one time-unit longer. $I_1 r$ represents this effect in the first round. When he makes his investment one time-unit earlier in the first round he has to do so in all future rounds too; these additional costs are included by the factor $1/(1-e^{-rT})$. Both kinds of additional capital costs add up to C_1. This first effect in condition (C11) is equal for every agent.

As a second effect shown in conditions (C11), agent i receives the benefits of the new technology instead of those of the old one. These benefits are the larger the later the agent's investment is due, *i.e.* the larger i is.

If condition (C11) is not satisfied, agent i benefits from advancing his investments by (at least) one time unit. Since agent T's incentives to make his investment earlier are greater than those of any other agent, it is sufficient to check whether condition (C11b) holds for agent T. So, the NE^{seq} pattern can only be an equilibrium if

(C12) $C_1 \geq a_1(T) - a_0(0)$

In other words, NE^{seq} can constitute an equilibrium only if the difference between $tech1$'s benefits with a full network and $tech0$'s stand-alone benefits cannot compensate for the agents' commitments.

Comparison of conditions (C10c) und (C12)

A comparison of (C10c) und (C12) shows immediately that both conditions cannot hold simultaneously. Consequently, except for the case where both conditions hold with equality, NE^{rap} and NE^{seq} cannot be an equilibrium at the same time.

See Figure 24 for illustration. The t-axis measures the time and the "position" of the agents. The vertical axis measures the benefits (and costs) at time t. The curves C_1' and C_1'' give two levels of commitment. The transition benefit functions, $a_1(t)$ and $a_0(T-t)$, depict the benefits at t with sequential adoption. For convenience, these curves are linear in Figure 24. The figure further assumes that the stand-alone benefits are equally large and normalized to 0, *i.e.* $a_0^0 = a_1^0 = 0$. We keep this simplification for the subsequent figures. As long as the stand-alone values are equal and large enough to yield non-negative net-benefits for a round they can be ignored for our purposes. (Of course, if $C_i > a_i(T)$ they must be positive.)

Let C_1 equal C_1'. When date l has been reached, $tech1$'s benefits become greater than $tech0$'s. However, no agent switches because $C_1 > 0$. As soon as $t=q$ has been passed, the benefits of the new technology are larger than C_1. Yet since assets of the old technology still exhibit some benefits, none of the next agents advances his switch to $tech1$. Only from date $j > q > l$ onwards are the additional benefits with $tech1$ large enough to cause remaining agents to "prematurely" switch to $tech1$. Hence, with $C_1 = C_1'$, agents i, $i \in (j,T]$, do not follow the sequential adoption pattern. Only if C_1 is beyond $a_1(T)$, *e.g.* C_1'', will every agent wait for his investment to become due, and NE^{seq} may constitute an equilibrium.

However, NE^{rap} cannot exist then. If $C_1 = C_1''$ then it pays to stay with $tech0$ even if an agent the only one. If $C_1 = C_1'$, in contrast, every agent throws away his $tech0$'s asset rather than being the only one left using $tech0$. For later reference, we define the threshold of C_1 in correspondence with condition (C12) as C_1^H:

(DV) $C_1^H := a_1(T) - a_0(0)$

Figure 24
Existence of NE^{seq} and NE^{rap} depending on agents' commitment

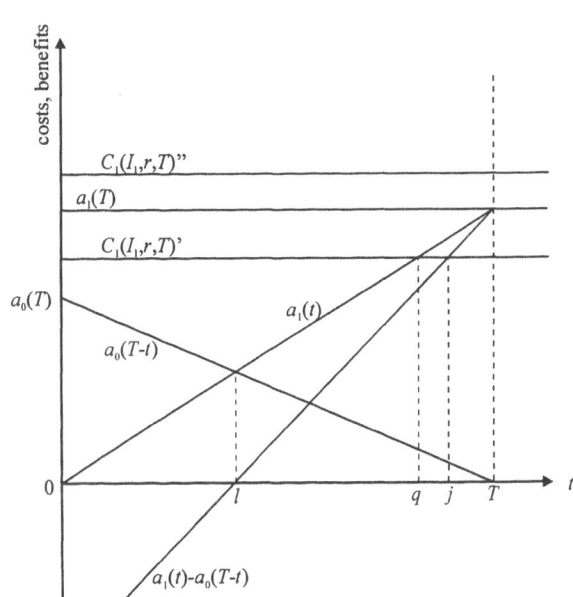

Incentives to postpone the switch

We now come to the second necessary condition for the existence of NE^{seq}, the "bandwagon condition". Remember, condition (C12) only insures that each agent i does not switch earlier than at $t=i$. But he might also have incentives to switch later. With condition (C12) being satisfied, agent 0's best alternative to be the first switcher is to be the last one (at $t=T+\varepsilon$). Agent 0 switches to *tech*1 at the beginning of the first round (at $t=0$) rather than at the beginning of the next round if

$$(C13) = BW^H \qquad -I_1 + \int_0^T a_1(t)e^{-rt}\,dt > -I_0 + \int_0^T a_0(T-t)e^{-rt}\,dt$$

For later reference we relable (C13) as BW^H. Condition BW^H says that agent 0's compounded *first round net benefits* with *tech*1 have to exceed those with *tech*0, given the sequential transition pattern. See Figure 25 for intuition.

$a_0(T-t)e^{-rt}$ and $a_1(t)e^{-rt}$ depict the transition benefit functions in discounted values. For simplicity, the figure assumed that $C_0 = C_1$ and, again, $a_0{}^0 = a_1{}^0 = 0$, implying that $a_0(T-t)e^{-rt} = 0$, $\forall t \geq T$. Figure 25 is similar to Figure 24, except that all values are *present* values (Figure 24 shows *actual* costs and benefits at t). The critical dates are unaffected, whether we use present or actual values.

Figure 25

Agent 0's alternative benefits facing sequential adoption

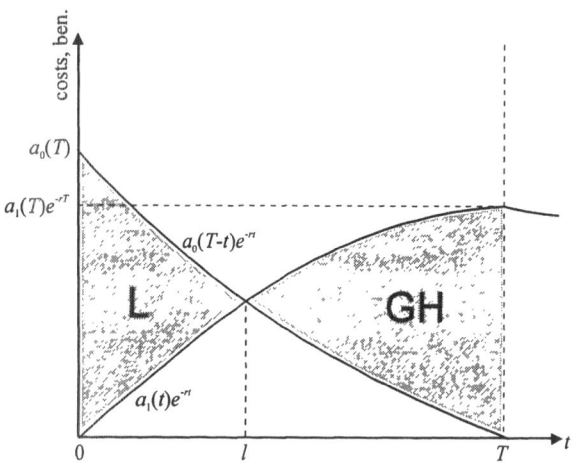

The areas under the curves represent the first round benefits with sequential adoption. Condition BW^H says that the area GH has to exceed the area L. When agent 0 invests in *tech*1 he will incur (relative) losses in the first part of the first round $[0,l)$. Later, within the interval $(l,T]$, he will be better off with *tech*1. Only when these later gains (area GH) exceed the early losses (area L), is agent 0 willing to switch to *tech*1 at $t=0$. In contrast, if these gains are smaller than the early losses, agent 0 adopts *tech*1 in the next round, *i.e.* just after $t=T$. Clearly, to wait until that date is agent 0's best alternative to a switch in $t=0$. Any later date cannot be profitable because at T the entire population adopts *tech*1. Any earlier date, *i.e.* $t \in (0,T]$, is worse than directly after T because sequential adoption requires $C_1 \geq C_1{}^H$.

The more time has passed the more agents have switched to *tech*1. Agent 0, therefore, has the strongest incentive to pursue a jump-on-the-bandwagon strategy. Thus, if condition BW^H holds for agent 0, an (analogous) condition also holds for any other agent.[172]

Hence, we can conclude: sequential adoption by the entire population is an equilibrium (NE^{seq}) if $C_1 \geq C_1^H$ and condition BW^H is satisfied.

Summary of the initial results

We have identified three (potential) Nash equilibria of the game. In the first one, NE^{est}, the entire population stays with the established technology perpetually. The second Nash equilibrium, NE^{rap}, is characterized by a "rapid" transition to *tech*1, *i.e.* the entire population switches to the challenging technology immediately after it has been introduced. Also in the third equilibrium, NE^{seq}, the new technology eventually prevails; the agents, however, switch "sequentially", *i.e.* the (complete) transition takes some time.[173]

The level of the agents' commitments to their adopted technology is crucial for the existence of these equilibria. The relationship between the marginal costs of capital (C_1) and the difference of $a_1(T)$ and a_0^0, as given in condition (C12) or (DV), defines the threshold for the possible existence of NE^{rap} and NE^{seq}. Only if we have relatively low commitments (marginal costs of capital), *i.e.* $C_1 < C_1^H$, does NE^{rap} exist. If, instead, agents face relatively large commitments, *i.e.* $C_1 \geq C_1^H$, NE^{seq} *might* exist, however its existence requires that condition BW^H is satisfied as well. Notice that the level of C_1 is only relevant for the network dependent part of the benefits. Thus, for example, provided that the full-network benefits remain unchanged, the "relative" level of C_1 is higher if the stand-alone benefits become larger.

See Table 5 for a summary.

---

[172] Note, as opposed to all former conditions, whether condition BW^H holds or not depends on the actual shape of $a_0(t)$ and $a_1(t)$.

[173] Precisely, the transition takes a time-period of T.

Table 5

Partial overview of potential equilibria

Equilibrium Pattern	Sufficient conditions
NE^{rap}	$C_1 < C_1^H$
NE^{seq}	$C_1 \geq C_1^H$ *and* condition BW^H holds
NE^{est}	Always exists

5.4.2 Equilibria With Strong Commitments $(C_1 \geq C_1^H)$

Our further analysis of equilibria is divided into two subsections according to the strength of the agents' commitments. In the *next* section (5.4.3), commitments are "weak", *i.e.* $C_1 < C_1^H$, so that both NE^{rap} and NE^{est} exist. In *this* section, commitments are "strong" $(C_1 \geq C_1^H)$, *i.e.* NE^{est} exists and, possibly, NE^{seq} alongside it.

In this section, we first demonstrate that strong commitments imply that no equilibrium (path) other than NE^{est} and NE^{seq} exists. Afterwards, it is demonstrated that for each parameter constellation, there exists a unique subgame perfect equilibrium.

Proposition 5.1: Let $C_1 \geq C_1^H$. No equilibrium (adoption paths) other than NE^{est} and NE^{seq} exist.

Sketch of proof: First, we argue that, in equilibrium, transition either starts at $t=0$ or never. Afterwards we demonstrate that equilibrium transition will either be sequentially or does not occur in the first place. Consequently, only NE^{seq} or NE^{est} can be equilibrium paths.

One might think that a continuum of equilibria exists with respect to the dates where adoption starts, *i.e.* sequential transition could start at some $t' \in (0,\infty)$, *e.g.* in $t'=T/2$. This is, however, not an equilibrium path because the agent who is located just before agent t' would switch before t'. Agent t' would only switch at t' if he is better off switching at $t=t'$ than at $t=T+t'$. Due to symmetry this, however, implies that agent $t'-\varepsilon$ is better off switching at $t=t'-\varepsilon$ than at $t=t'-\varepsilon+T$. This is true for any $t' \in (0,\infty)$. Thus, in equilibrium, adoption can only start at $t=0$.

Within an equilibrium path all agents adopt either *tech*0 in any $t \in [0,\infty)$ or agents switch to *tech*1 sequentially during the interval $[0,T]$. Imagine there were some discontinuity during sequential adoption at some $t' \in [0,T]$, *e.g.* at $t'=T/2$. The same argument as above suggests that this is not possible in equilibrium. □

Note that, strictly speaking, our equilibria are equilibrium *paths* because they can be supported by more than one strategy profiles. Since agents know the state at any t, they may play strategies that entail conditioning of actions on history (closed-loop strategies). Hence, there are many strategies supporting these paths. For instance, strategies of agent $i{\neq}0$ supporting NE^{seq} are "switch if agent 0 has switched", or "switch if all your predecessors have switched", etc. *Proposition* 5.1 shows only that no equilibrium *path* other than NE^{est} and NE^{seq} can be supported by a strategy profile.[174] Only if we make further assumptions on how agents form their strategies, each path could have a unique strategy profile. For example, one could assume that agents include the actions of any subset of agents at any t into their conditions.

We continue our analysis as follows. First, the agents' transition gains and losses will be compared. Second, a condition is established under which agent 0 benefits from transition to *tech*1. Finally, it will be shown that NE^{seq} is the unique subgame perfect equilibrium path if this condition holds. If it does not hold, the population stays with *tech*0.

Gains and losses through transition

To select between NE^{est} and NE^{seq} we need to compare the agents' transition benefits vis a vis their position.

In Figure 26, which derives from Figure 25 above, we see that agent l benefits most from the transition to *tech*1 because he always applies that technology which exhibits the most benefits. Agent b gains as much as agent 0 does: the amount that agent 0 (comparably) loses in the early stages $t=[0,l)$, represented by the area L, equals exactly the amount that agent b loses from $t=l$ to $t=b$, represented by L'. Every agent situated "behind" agent b is worse off through transition than agent 0. If agent 0 marginally favored sequential transition over a stay with *tech*0, then each agents i, $i \in (b, T]$, would prefer that the population stayed with *tech*0. In any case, agent T is the one who gains least (loses most) through transition to *tech*1. Certainly, he must always be worse off than agent 0, otherwise agent 0 would rather switch to *tech*1 at $t=T$ (*i.e.* condition BW^H fails).

It follows that agent 0 might favor sequential transition while there are agents who want that the population stays with *tech*0. For those agents, transition costs, produced by the period of

---

[174] Recall that we excluded mixed strategies.

incompatibility, are too high. However, if agent 0 prefers transition to perpetual stay, then each agent i does so *once t=i is reached.*

Figure 26

Gains and losses through transition

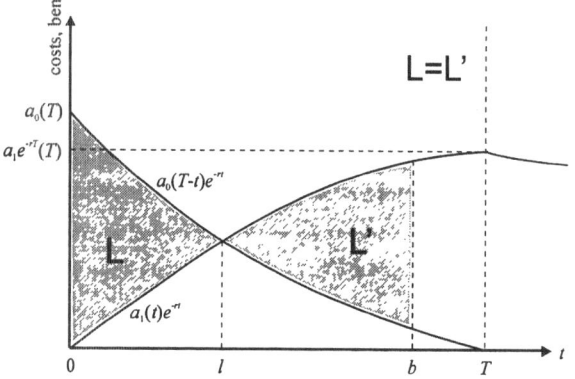

Agent 0 prefers sequential transition to perpetual stay, iff

$$\text{(C14a)} \quad \int_0^T a_1(t)e^{-rt}\, dt - \int_0^T a_0(T)e^{-rt}\, dt + \frac{a_1(T)-a_0(T)}{r}e^{-rT} - (I_1 - I_0)\frac{1}{1-e^{-rT}} \geq 0$$

Rearranging yields

$$\text{(C14b)} = \text{PR}^{\text{H}} \quad \int_0^T a_1(t)e^{-rt}\, dt - \int_0^T a_0(T)e^{-rt}\, dt + \frac{[a_1(T)-a_0(T)]e^{-rT} - [C_1 - C_0]}{r} \geq 0$$

For later reference we rename condition (C14b) as PR^{H}, the "preference" condition. In words: the difference of the marginal capital costs $(C_1 - C_0)$ is constant from $t=0$ onwards. From $t=T$ onwards, in both alternative paths, the benefits stay constant too, as every agent adopts *tech*1. Since *tech*1 is superior to *tech*0, all agents enjoy larger payoffs from $t=T$ onwards. In the first round, however, agents might be worse off through transition. If agent 0's (potential) first round losses are not compensated through the gains in subsequent rounds, condition PR^{H} does not hold.

Figure 27

Agent 0's gains and losses with transition to tech1

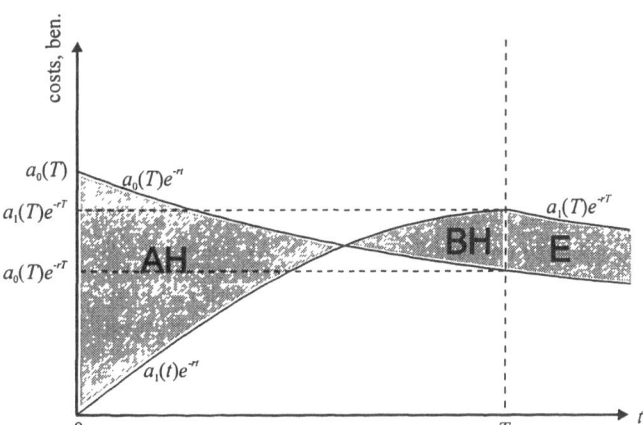

Figure 27 illustrates.[175] $a_0(T)e^{-rt}$ gives the present values of benefits at t when all agents stay with *tech0*. $a_1(t)e^{-rt}$ depicts the transition benefit function (in compounded values). If area AH is smaller than BH + E, then agent 0 prefers transition (PR^H holds). If not, agent 0 favors a perpetual stay with the established technology.

These facts are used to identify a unique subgame perfect equilibrium:

Proposition 5.2: Let $C_1 \geq C_1^H$ and suppose that BW^H holds (*i.e.* NE^{seq} and NE^{est} exist). If condition PR^H holds, NE^{seq} is the unique subgame perfect equilibrium. If PR^H fails, NE^{est} is the unique subgame perfect equilibrium.

Sketch of proof: First suppose $C_0 \geq C_1 \geq C_1^H$. This implies that we can ignore agents' option to switch back to *tech0* in the first round once they have switched to *tech1*. Consequently, the game is "sequential", where at any t only agent $i=t$ makes a technology choice.[176] Although the game's horizon is infinite, we can apply backward induction because sequential adoption of *tech1* leads to a steady state once $t=T$ is reached where all agents adopt *tech1* forever.[177]

---

[175]   In Figure 27, we again assume that $I_0 = I_1$ and $a_0^0 = a_1^0 = 0$.

[176]   As far as the term "sequential" applies to continuous time games.

[177]   See also Regibeau and Rocket (1996).

Building upon Farrell and Saloner (1985)[178], a sufficient condition for transition to be the unique subgame perfect equilibrium is

(C15)
$$\int_i^T a_1(t)e^{-rt}\,dt + \frac{a_1(T)e^{-rT} - C_1}{r} \geq$$
$$\int_i^T a_0(T-t)e^{-rt}\,dt + \int_T^{T+i} a_0(t-i)e^{-rt}\,dt + \frac{a_0(T)e^{-r(T+i)} - C_0}{r} \qquad , \forall i$$

(C15) insures that, at any $t=i$, agent i prefers transition to $tech1$ to staying with $tech0$, whatever his beliefs about future adoption behavior. (C15) assumes the most favorable adoption path for $tech0$ (transition back to $tech0$ as fast as possible).

For agent 0 condition (C15) is equivalent to PR^H. Intuitively, if PR^H is satisfied, (C15) is also satisfied for each agent i. Thus, backward induction selects NE^{seq} if PR^H holds. If (C15) holds, agent T's predecessor knows that agent T switches if all agents before him have switched. This is anticipated by the predecessor of agent T's predecessor, and so forth. (Once $t=T$ is reached, or even before $t=T$, transition back to $tech0$ cannot occur in equilibrium because the analogous condition to BW^H cannot hold.)

If PR^H fails, the population stays with $tech0$. If agent 0 prefers that the population perpetually stays with $tech0$ and does not switch, his successor faces exactly the same situation that agent 0 faced. Thus, agent 0's successor will not switch and so his successor will also not switch, etc.

Now suppose $C_1 \geq C_1^H \geq C_0$. The difference to the case $C_0 \geq C_1 \geq C_1^H$ is that it is not necessarily a dominated strategy to switch back to $tech0$. In each subgame at $t=i$, those agents who have already switched (agents h, $h<i$) and agent i simultaneously decide whether to stay with $tech1$ (agents h) and to switch (agent i), respectively. In order to apply backward induction, we make the following assumption:

Assumption (5.C): In any subgame starting at $t=i$, agents h ($h<i$) in conjunction with agent i are able to coordinate on their "best" equilibrium (anticipating adoption choices of agents m, m>i.).[179]

Condition (C15) becomes

---

[178] See the brief discussion of that model in section 5.2.1.

[179] We briefly illustrate this assumption below. Assumption (5.C) amounts to the network effects literature standards assumption as discussed in subsection 5.2.2. The assumption also insures that transition back to $tech0$ never occurs once all agents adopt $tech1$.

$$(C16) \quad \int_{i}^{T} a_1(t)e^{-rt}\,dt + \frac{a_1(T)e^{-rT} - C_1}{r} \ge \int_{i}^{T} a_0(T)e^{-rt}\,dt + \frac{a_0(T)e^{-rT} - C_0}{r} \quad ,\forall i$$

Again, (C16) insures that each agent i at $t=i$ prefers transition to $tech1$ to staying with $tech0$, *whatever* his beliefs about future adoption. (C16) assumes the most favorable adoption path for $tech0$ (full network for $tech0$ from $t=i$ onwards).

Again, for agent 0, condition (C16) is equivalent to PR^H. If PR^H is satisfied, (C16) is satisfied for any agent i. Thus, backward induction selects NE^{seq} if PR^H holds.

Analogously to above, if PR^H fails the population stays with $tech0$. □

Illustration of Assumption (5.C)

As already mentioned in section 5.2.1, to make Assumption (5.C) or a similar one has become standard in the network effects literature. For illustration of this assumption consider a simple 2-stage game as displayed in Figure 28. Without imposing the restrictions of Assumption (5.C), the game has two subgame-perfect (pure-strategy) equilibria, (l_1,l_2) and (r_1,r_2,r_2). In applying Assumption (5.C), however, backward induction selects (r_1,r_2,r_2). Since (r_2,r_2) is the best equilibrium in stage 2, player 1 plays (r_1,r_2), as he knows that player 2 will play r_2.[180]

Figure 28

Illustration of Assumption (5.C)

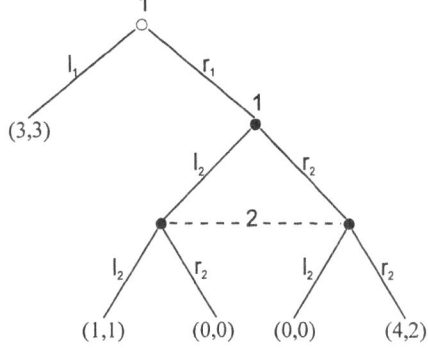

---

[180] The method of Forward Induction would select the same equilibrium, too.

Summary

To conclude this subsection, if $C_1 \geq C_1^H$, there is always a unique subgame-perfect equilibrium. Depending on the parameters $(a_0(t), a_1(t), r, T, I_1, I_0)$, either sequential adoption occurs (BW^H and PR^H hold) or the population stays with the established technology "forever" (BW^H or PR^H or both fail).

A comparison of conditions BW^H and PR^H yields that both paths NE^{est} and NE^{seq} can actually result. Whether agent 0 favors the adoption path according to NE^{seq} or to NE^{est} and whether BW^H holds or not, cannot be said in general, as we do not specify the benefit functions. It might very well happen that BW^H holds while PR^H fails, and vice versa – even with equal investments and stand-alone values. See section 5.5 for implications.

5.4.3 Equilibria With Weak Commitments $(C_1 < C_1^H)$

We now turn to low levels of commitment, *i.e.* $C_1 < C_1^H$. Remember that C_1 represents agents' commitments. It indicates how much *tech*1's benefits have to exceed *tech*0's in order to cause agents to switch to *tech*1 even though their *tech*0-asset is still in working order. If $C_1 \geq C_1^H$, as assumed in the previous subsection, agents' optimal choice is always to wait until their *tech*0-asset's life expires. However, when $C_1 < C_1^H$, which is what we assume now, agents would switch earlier if *tech*1's network is sufficiently larger than *tech*0's. As we have seen in subsection 5.4.1, this causes (at a minimum) NE^{rap} to coexist with NE^{est}. Moreover, as we shall see, further equilibria may exist alongside it.

It is first demonstrated that $C_1 < C_1^H$ implies that a continuum of equilibria exists that start with sequential adoption in $t=0$ followed by a simultaneous switch of all "remaining" agents within some later defined interval. Afterwards, we show that no other equilibrium path exists beside those within this continuum. Next, it is argued that there are only two "plausible" equilibrium paths within this continuum. Finally (in a way similar to the one taken in the former section 5.4.2) it is shown that for each parameter constellation, a unique "plausible" equilibrium path exists.

The "NE $^{k\text{-}rap}$ continuum"

Define date j as

(DVI) $$j := t \mid C_1 = a_1(t) - a_0(T - t)$$

Date j has already appeared in Figure 24. Given agents switch sequentially to tech1, $t=j$ is the critical date at which the number of *tech*1 users is so large that the remaining *tech*0 users prefer to join *tech*1, thereby throwing away their (still working) *tech*0 asset. In other words, $t=j$ is the date at which the benefits of the diffusing technology are large enough in relation to those of the preestablished technology to compensate for the agents' commitments to *tech*0.

Since $C<C_1^H$, any path starting with sequential adoption at $t=0$ and entailing a simultaneous switch at $t=k$, $k \in [0,j]$, may constitute a Nash equilibrium. (This set also contains NE^{rap}). Denote the set of such paths as the "$NE^{k\text{-}rap}$ continuum". A particular path is referred to as $NE^{k\text{-}rap}$ where k refers to the date where the simultaneous switch occurs.

That such paths might be equilibrium follows from the existence of NE^{rap}. No agent i, $i \geq k$, has an incentive to make his investment later than at $t=k$ because $a_1(T)-a_0^0 > C_1$, and also not earlier because $a_1(t)-a_0(T-t) < C_1$, $\forall t<k$.

Figure 29
Benefits within the NE $^{k\text{-}rap}$ continuum

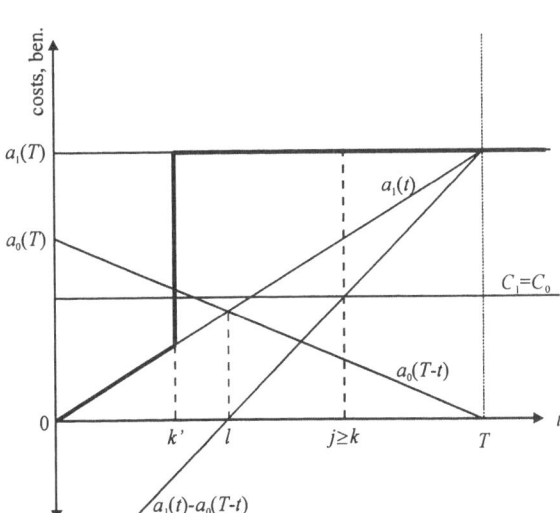

In Figure 29, the striped area indicates *tech*1's benefits within the $NE^{k\text{-}rap}$ continuum.[181] The thick line gives the benefits with $NE^{k'\text{-}rap}$ as an example.

---

[181] In Figure 29, $I_0=I_1$ and $a_0^0 = a_1^0$.

Proposition 5.3: Let $C_1 < C_1^H$. Then, no other equilibrium path other than NE^{est} and those included in the NE $^{k\text{-}rap}$ continuum exists.

Sketch of proof: The proof is similar to that of *Proposition 5.1*. To prove the proposition it must first be shown that no transition path can start at any $t\neq0$. Next, it is to be shown that no equilibrium path can involve any discontinuity other than at $t=k$. Thus, we can refer to *Proposition 5.1*. □

In order to have NE $^{k\text{-}rap}$ actually exist, no agent may have an incentive to jump on the bandwagon, *i.e.* to reinvest in *tech*0 at $t=i$ and switch to *tech*1 at $t=k$. That is to say, an analogous condition to condition BW^H must be satisfied. Since we later show that we can rule out any equilibrium except NE^{est} and NE $^{j\text{-}rap}$, we focus on the condition that establishes NE $^{j\text{-}rap}$.

Provided that agent 0 expects an adoption path described by NE $^{j\text{-}rap}$, his best alternative to a switch at $t=0$ is a switch at $t=j$. Agent 0 switches at $t=0$ rather than at $t=j$ iff

$$(\text{C17a}) \qquad \int_0^j a_1(t)e^{-rt}\, dt - I_1 \geq \int_0^j a_0(T-t)e^{-rt}\, dt - I_0 - I_1 \frac{e^{-rj} - e^{-rT}}{1 - e^{-rT}}$$

Rearrangements lead to

$$(\text{C17b}) = \text{BW}^L \qquad \int_0^j a_1(t)e^{-rt}\, dt - I_1 \frac{1 - e^{-rj}}{1 - e^{-rT}} \geq \int_0^j a_0(T-t)e^{-rt}\, dt - I_0$$

For later reference (C17b) is relabeled as BW^L. BW^L amounts to the bandwagon condition for low levels of commitment. On the *lhs* of condition (C17a) one finds the net benefits of agent 0 when switching at $t=0$, and on the *rhs* his net benefits if he adopts *tech*0 at $t=0$ again and switches to *tech*1 only at $t=j$. The last term on the *rhs* represents the costs of switching T-j time-units earlier than at the beginning of the second round. The benefits from $t=j$ onwards are omitted, as they are equal in both alternatives (*i.e.*, $a_1(T)e^{-rt}$). Note, that the second term on the *lhs* in condition BW^L is greater than -I_1.

Analogous to condition BW^H, incentives to jump on the bandwagon are larger for agent 0 than for any agent located behind him. Thus, if condition BW^L holds, no agent has an incentive to postpone his switch. If BW^L is not satisfied, agent 0 would adopt *tech*0 at $t=0$ and switch to *tech*1 later. If this is true for agent 0, it will also be so for agent 0's successor (due to symmetry) and thus also for that one's successor, and so forth, implying that NE $^{j\text{-}rap}$ is not an equilibrium.

Figure 30 illustrates condition BW^L.[182] If agent 0 switches at $t=j$ rather than at $t=0$, he receives additional benefits represented by the area L. Between $t=l$ and $t=j$ he sacrifices benefits, as during this period, he is stuck with the *tech0* although *tech1* already exhibits more benefits. In addition, he has to incur additional investment expenses. For $I_0 = I_1$, these amount to

(E20) $$I_1 \frac{e^{-rT} - e^{-rj}}{1 - e^{-rT}} \quad \Leftrightarrow \quad \int_j^T C_1 e^{-rt} dt$$

which are represented by the area GL$_2$.

Figure 30
Agent 0's alternative benefits facing NE $^{j\text{-rap}}$

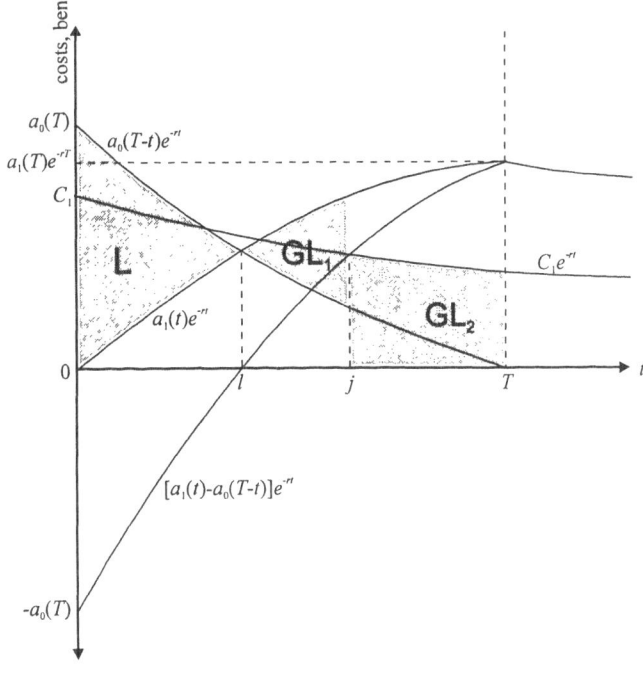

---

182   Again, in Figure 30, $I_0 = I_1$, $a_0^0 = a_1^0 = 0$.

If area L is smaller than $GL_1 + GL_2$, then $NE^{j\text{-}rap}$ exists. The area L is the same for both strong and weak commitments. GL_2 distinguishes BW^L from BW^H (see Figure 25 for comparison). The areas GL_1 and GL_2 in Figure 30 supplant the area GH in Figure 25. We shall later show that $GL_1 + GL_2$ is always smaller than GH.

Equilibrium Selection

We argue now that $NE^{j\text{-}rap}$ and NE^{est} are the only "plausible" equilibrium paths supportable by closed-loop strategies. Since agents find themselves in an environment where each agent's action set is non-trivial – implying some uncertainty regarding the states at t – every agent will find it beneficial to condition his actions at t on the history of t. That is, agents adopt closed-loop strategies. The reasoning that rules out every equilibrium transition path but $NE^{j\text{-}rap}$ is as follows. Agents have an incentive to behave apathetically. No agent whose asset is still in working order would risk switching to *tech*1 before its benefits are sufficiently larger than *tech*0's. Each agent is free to choose the time when to procure the new technology's asset, and thus he can move *immediately* to *tech*1 once a sufficient number of agents have done so – why take the risk of getting stuck with the new technology too early? Hence, no agent should expect that any agent $i<j$ switches before $t=i$ and any agent $i≥j$ switches before $t=j$.

Even if there is some time lag, it seems to be a reasonable assumption that the risk of adopting the new technology too early is often too high. Nevertheless, there are environments that do not fit our assumption. For instance, a switch to another production technology may entail firm-specific R&D and other preparation that may take a not inconsiderable amount of time. If such actions are unobservable by other firms, then these cannot *follow* immediately. We briefly discussed this issue in subsection 2.2. If such a period is long enough, the game may be similar to a simple (simultaneous) coordination game such as the one displayed in Figure 2.

More generally, our assumption especially applies to markets where the time lag between *unobservable* adoption decision and *observable* actual adoption does not materially exist. Moreover, the set of players should be sufficiently large. Otherwise one may expect that they are able to coordinate their actions. Thus, our setup applies specifically to typical widely adopted consumer goods. There is not much preparation to be done before buying such goods. Consumers can go and get a new CD-player or TV set at any time they wish. If so, there is no reason why a consumer should switch to the new technology unless it already makes him better off.

We now briefly give a theoretical justification for this argument. Since a single agent does not possess "strategic power", *i.e.* he cannot trigger other agents' actions, his best choice is to adopt the payoff maximizing technology at any t, unless his asset has to be renewed. For instance, in $t=0$, where the new technology has just become available, each agent i, $i>0$, has the following options: he can either throw away his working asset and switch now at $t=0$ to $tech1$ or he can stay with $tech0$. The alternative payoffs depends on the technology he adopts (*i.e.* $a_0(x)$ or $a_1(x)$), and on how many other agents adopt the same technology (*i.e.* x). With the first option, agent i receives $a_1(x) - C_1$, and with the second option, *i.e.* staying with $tech0$, he gets $a_0(x)$. The minimum payoff with the one option is always smaller than the maximum payoff with the other option. This is a corollary of the existence of NE^{rap}. Consequently, there would be two "static" pure strategy equilibria at $t=0$: either the entire population switches to $tech1$ or stays with $tech0$ (except agent 0).

However, any agent i, $i > 0$, knows that he could immediately "follow" to $tech1$, if a sufficient number of agents have switched at $t=0$. Consequently, the maximum relative losses pertaining to the option of staying with $tech0$ (in comparison to a switch to $tech1$) approach zero. Switching to $tech1$ in $t=0$, however, *could* cause positive losses since the investment into $tech1$ is sunk. Thus, any agent i, $i>0$, faces at $t=0$ a decision problem as illustrated by the matrix in Figure 31.

Figure 31
Decision problem of agent k>i at t=i

Other agents

		S_ϕ	S_ρ
agent i	S_ϕ	B	A
	S_ρ	Ω	A

$B \lessgtr A$, $\Omega < \min(A, B)$

The row player represents some agent $i>0$, and the column player summarizes all remaining agents (except agent 0). The payoffs within the matrix apply for "both" players. S_ϕ denotes the strategy "wait and follow" and S_ρ is "switch to $tech1$". With (S_ρ, S_ρ), both players switch (*i.e.* all agents switch to $tech1$ immediately after it has been introduced), and receive a payoff A. In (S_ϕ, S_ρ), agent i still adopts $tech0$ at $t=0$; however since he can follow to $tech1$ without a time lag, his payoff with this constellation goes towards A. With both players playing S_ϕ, agent i gets B. Depending on agent i's position and expectations for the future game, B can be smaller or greater than A (early agents tend to prefer a rather early simultaneous switch and later agents a rather late one). However, B is always greater than Ω which is what he receives in (S_ρ, S_ϕ). If he switches to *tech*1 at $t=0$ while all other agents stay with *tech*0, he is definitely on a lower (gross) benefit level than with *tech*0. In addition he has costs of C_1. Even if the entire population switches immediately afterwards, he would not have done worse with S_ϕ. And the probability of this event is definitely smaller than 1. Thus, $B > \Omega$, and, therefore, S_ϕ weakly dominates S_ρ. Hence, (S_ϕ, S_ϕ) remains the only trembling hand perfect equilibrium.

Therefore, agents do not expect any other agent i to switch at $t=0$, and so no agent i switches at $t=0$. Since no single agent's decision materially affects other agents' payoffs, there is no other strategy available that is not (weakly) dominated by S_ϕ. Thus, it is rational for those agents to stay in apathy.

This game is played at each t by all agents whose investment has not fallen due, as long as the number of *tech*1 users is smaller than j. If j or more agents use *tech*1, Ω becomes greater than B (following from Definition (DVI)). (S_ρ, S_ρ) is the only equilibrium (by dominated strategies), implying that the entire population will adopt *tech*1.[183]

Thus, we can set up the following proposition:

Proposition 5.4: Let $C_1 < C_1^H$. In equilibrium entailing transition to *tech*1, no agent i makes his investment earlier than at $t=i$, unless j agents apply *tech*1.

Following from *Proposition 5.3* and *Proposition 5.4*, two equilibrium paths remain, NE$^{J\text{-}rap}$ and NE^{est}. Now the condition is established under which agent 0 favors transition according to NE$^{J\text{-}rap}$ over NE^{est}. Afterwards, it is shown that a unique subgame perfect equilibrium exists. Agent 0 favors a path according to NE$^{J\text{-}rap}$ over NE^{est} if

$$\text{(C18a)} \quad \int_0^J a_1(t)e^{-rt}\,dt + \int_J^T a_1(T)e^{-rt}\,dt + \frac{a_1(T)-a_0(T)}{r}e^{-rT} - (I_1-I_0)\frac{1}{1-e^{-rT}} \geq 0$$

Incorporating (DIV) yields

$$\text{(C18b)} = PR^L$$

$$\int_0^J a_1(t)e^{-rt}\,dt + \int_J^T a_1(T)e^{-rt}\,dt - \int_0^T a_0(T)e^{-rt}\,dt + \frac{[a_1(T)-a_0(T)]e^{-rT} - [C_1-C_0]}{r} \geq 0$$

---

[183] As mentioned in section 5.2.1, our method to select among equilibria within the NE $^{k\text{-}rap}$ continuum is similar, however not equivalent, to Cabral (1990). Cabral's method would select the same path in our framework.

Condition PR^L (relabeled from (C18b) for later reference) is derived analogous to condition PR^H. Not surprisingly, a comparison of these conditions reveals that condition PR^L is more likely to be satisfied than condition PR^H. The reason is that *tech1* obtains its full network earlier. See Figure 32 for illustration.

Figure 32
Agent 0's gains and losses with NE $^{j\text{-}rap}$

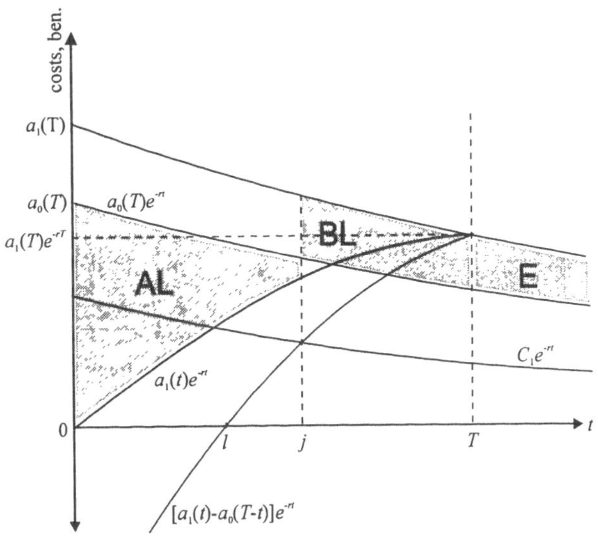

In Figure 32, the areas AL and BL supplant the areas AH and BH, respectively, from Figure 27 If the BL + E > AL, agent 0 prefers NE$^{j\text{-}rap}$ to NE^{est}. BL is always greater than BH and AL is never greater than AH, which implies that it is more likely that agent 0 prefers transition when the commitments become weaker.

Proposition 5.5: Suppose that BW^L holds (*i.e.* NE$^{j\text{-}rap}$ and NE^{est} exist). If condition PR^L holds, NE$^{j\text{-}rap}$ is the unique subgame perfect equilibrium. If PR^L fails, the entire population perpetually stays with *tech0*.

Sketch of proof: The proof is analogous to that of *Proposition* 5.2 for the case $C_1 > C_1^H > C_0$. Therefore, we only set up the analogous condition and refer to *Proposition* 5.2.

Backward induction now starts from *t=j*. The analogous condition to (C16) is

(C19)
$$\int_i^j a_1(t)e^{-rt}dt + \int_j^T a_0(T)e^{-rt}\,dt + \frac{a_1(T)e^{-rT}-C_1}{r} \geq$$
$$\int_i^T a_0(T)e^{-rt}\,dt + \frac{a_0(T)e^{-rT}-C_0}{r} \qquad ,\forall i < j$$

$$\int_i^T a_0(T)e^{-rt}\,dt + \frac{a_1(T)e^{-rT}-C_1}{r} \geq$$
$$\int_i^T a_0(T)e^{-rt}\,dt + \frac{a_0(T)e^{-rT}-C_0}{r} \qquad ,\forall i \geq j$$

(C19) insures that each agent i at $t=i$ prefers transition to *tech*1 to staying with *tech*0, *whatever* his beliefs about future adoption. (C19) assumes the most favorable adoption path for *tech*0 (full network for *tech*0 from $t=i$ onwards).

Again, both conditions are definitely satisfied if PR^L is. □

To conclude, also with relatively low levels of commitment $(C_1 < C_1^H)$, there is always a unique "plausible" equilibrium path. Depending on the parameters $(a_0(t), a_1(t), r, T, I_1, I_0)$, either transition occurs or the population stays with the old technology perpetually. If transition occurs, then it takes a discontinuous path.

5.4.4 Summary of Plausible Nash Equilibria

Our first result is that new superior network technologies have a chance to become adopted even in a general setting where the population is large, first adopters of the new technology are worse off at the beginning, their choice does not relevantly affect the payoffs of other players, and (if $C_1 \leq C_1^H$) players can switch between technologies at any time. Moreover, within the framework of our model, we can predict whether and, if so, how long the transition takes. Both the transition time (τ) and "whether or not" depend on various parameters. Figure 33 gives an overview of the partition of plausible equilibria.

Necessary and sufficient for transition to *tech*1 is that two conditions are satisfied. First, agent 0 must benefit from being the first adopter rather than the last one (condition BW^H or BW^L, respectively). That is to say, agent 0, *when expecting transition to tech*1, must not have an incentive to adopt *tech*0 now and switch to *tech*1 later. In other words, "early" agents may not have an incentive to pursue a jump-on-the-bandwagon strategy. As a second condition, agent 0 must prefer transition to the new technology in comparison to "perpetually stay" with the established one (condition PR^H or PR^L, respectively). If agents' commitments are strong

$(C_1 \geq C_1^H)$, conditions BW^H and PR^H are relevant, otherwise BW^L and PR^L are. If one or both conditions fail, the population stays with the established technology.

Figure 33
Equilibrium partition

Equilibrium Pattern	Sufficient conditions
$NE^{j\text{-}rap}$	$C_1 < C_1^H$, BW^L *and* PR^L hold
NE^{seq}	$C_1 \geq C_1^H$, BW^H *and* PR^H hold
NE^{est}	Otherwise

The transition path always starts with "sequential" adoption immediately after the new technology has been introduced. If the commitments are strong, then the entire population adopts *tech*1 "sequentially" and the transition time (τ) is equal to the assets' life span T. If the commitments are weak, then we obtain a discontinuous transition path: at date $j<T$, all remaining agents switch simultaneously, which implies that $\tau=j$. The lower the level of the agents' commitments the earlier is the j-date and, thus, the shorter is the transition time.

As discussed in section 2.6, such discontinuous adoption paths are consistent with many empirical findings. Of course, in reality such paths do not involve a *perfectly simultaneous* switch of the remaining agents at some critical date. Our framework would produce more "smooth" discontinuities if we allowed for some heterogeneity of agents' preferences and some delay in the diffusion of information, without changing the basic results.[184]

5.5 Welfare Analysis – Commitments and Efficiency of Transition

Transition always includes a period of temporary incompatibility. Due to these (opportunity) costs, transition is not always desirable even though the challenging *tech*1 is "statically" superior to the established *tech*0. In section 5.4, it has been demonstrated that whether or not transition occurs depends on two conditions. First, agent 0 has to favor transition over the perpetual stay. Second, agent 0, even when expecting transition, must not have an incentive to pursue a jump-on-the-bandwagon strategy. Since, thus, whether or not transition occurs only depends on what is optimal for agent 0, one becomes skeptical whether the outcomes of the

---

[184] More heterogeneity would probably slow down the adoption process, as some early and interior agents would have incentives to postpone their switch. Of course, too much heterogeneity and too much information delay may change our results (as can be derived from our analysis in section 3).

game are optimal from a societal point of view. Does the population suffer from excess momentum and/or excess inertia? Furthermore, each agent contemplating the timing of his switch takes into account only his benefits. Do the agents' timing decisions produce an optimal transition path (including an optimal transition time)?

In order to maintain the most possible generality of the results, the transition benefit functions $(a_i(t))$ will stay unspecified even within the welfare analysis. However, we use linear network benefit functions (as introduced in section 5.3) as a reference case. To make our points clear, it makes sense to impose some further simplifications, which are summarized in Assumption (5.D)

Assumption (5.D) $I := I_0 = I_1$

$$a^0 := a_0^0 = a_1^0$$

$$f(t) = 1/T$$

Assumption (5.D) states that the investments required to adopt both technologies, as well as the stand-alone benefits of the technologies, are equal.[185] These simplifications have already been used in most of the above figures. Further, the assumption states that the agents' investment dates (agents' "positions") are equally distributed. That is to say, "at each t" the same "fraction" of agents renews its assets. This assumption makes probably most sense if the established technology has been in use for a long time. Of course, if $tech0$ has been established only for a short time and the level of the agents' commitment is low, then the distribution of the agents' positions is very likely not to be so smooth.

In order to compare the outcomes of the game with the socially optimal ones it is necessary to develop some further expressions. First, we need to know how the agents' positions affect the individual transition benefits. The transition benefits (gross of the regular investment expenses) of agent i are [186]

---

[185]   That the stand-alone values are equal while the network dependent part differs is consistent with the evolutionary model in section 3. It, however, departs from some established models, which assume quite the opposite. That is, some authors assume that the network dependent part of the benefit function is equal for both technologies while the stand-alone values differs. It certainly depends on the particular environment under consideration, which of the alternative assumptions is appropriate. For most of the aforementioned examples of network goods, the here chosen version seems more natural. See also Ellison and Fudenberg (2000).

[186]   An analogous expression applies to the case of strong commitments $(C_1 \geq C_1^H)$.

$$(E21) \quad u^{trans}(i) = \begin{cases} \int_0^i a_0(T-t)e^{-rt} + \int_i^j a_1(t)e^{-rt}dt + \int_j^T a_1(T)e^{-rt}\,dt + \dfrac{a_1(T)e^{-rT}}{r} & , for\ i < j < T \\[3mm] \int_0^j a_0(T-t)e^{-rt} + \int_j^T a_1(T)e^{-rt}dt + \dfrac{a_1(T)e^{-rT}}{r} - \int_j^i C_1 e^{-rt}dt & , for\ i \geq j < T \end{cases}$$

The (gross) benefits of agents if the population stays with the established *tech0* are equal for each agent and independent from C. These are

$$(E22) \qquad\qquad\qquad\qquad u^{stay} = \frac{a_0(T)}{r}$$

Thus, agent i's "transition surplus" is

$$(E23) \qquad\qquad\qquad\qquad w(i) = u^{trans}(i) - u^{stay}$$

Notice that condition PR^i, $i=\{H,L\}$, is equivalent to $w(0) \geq 0$. Due to our assumption that $I_0 = I_1$, condition BW^i, $i=\{H,L\}$, is equivalent to $u^{trans}(0) \geq u^{trans}(T)$. Thus condition BW^i is satisfied if agent 0's transition benefits are greater than those of agent T. If agent T's transition benefits exceeded agents 0's, agent 0 would simply take the position of agent T. In other words, the one who switches last must be worse off than the one who switches first. Thus, in order to check whether condition BW^i holds, we only need to compare the transition benefits of agent 0 and agent T. Since the agents' benefits with perpetual stay are independent of their position, $u^{trans}(0) \geq u^{trans}(T)$ also implies that $w(0) \geq w(T)$.

(E21) indicates that the agents' benefits only differ throughout the first round, *i.e.* during the interval $t=[0,T]$. For later reference denote these first round transition benefits of agent i as

$$(E24) \quad u^T(i) = \begin{cases} \int_0^i a_0(T-t)e^{-rt} + \int_i^j a_1(t)e^{-rt}dt + \int_j^T a_1(T)e^{-rt}\,dt & , for\ i < j \\[3mm] \int_0^j a_0(T-t)e^{-rt} + \int_j^T a_1(T)e^{-rt}dt - \int_j^i C_1 e^{-rt}dt & , for\ i \geq j \end{cases}$$

As an indicator for utilitarian social welfare we use agents' average benefits. These are assumed to be equal to the expected benefits that an agent would receive if he did not know his position in advance and each position that will be assigned to him was equally likely. Thus, transition and perpetual stay, respectively, produce average benefits of:

$$(E25) \qquad U^{trans} = \frac{1}{T} \left(\int_0^T u^T(i)di \right) + \frac{a_1(T)e^{-rT}}{r}$$

$$(E26) \qquad U^{stay} = \frac{a_0(T)}{r}$$

Finally, the average surplus of transition is

$$(E27) \qquad W = U^{trans} - U^{stay}$$

Thus, transition is desirable if

$(C20) = SW^i (i=\{H,L\})$ $\qquad\qquad W \ge 0$

See Table 6 for summary.

Table 6

Agent 0's and social willingness to switch

	Sufficient conditions
Transition occurs	BW^i *and* PR^i ($i \in \{H,L\}$) hold, *i.e.* $w(0) \ge w(T)$ *and* $w(0) \ge 0$
Transition is desirable	SW^i ($i \in \{H,L\}$) holds, *i.e.* $W \ge 0$

The remainder of section 5.5 is structured as follows. We first analyze the risk of excess momentum and excess inertia in our framework. In subsection 5.5.1, we do so for the case of strong commitments. In subsection 5.5.2, we do the same for the case of weak commitments. Section 5.5.3 summarizes the results obtained so far. Afterwards, in section 5.5.4, we deal with the transition time.

5.5.1 Strong Commitments: Excess Momentum and Excess Inertia

We first show that strong commitments imply that both excess momentum and excess inertia could occur. Afterwards, we look at both scenarios in more detail. We find that if network benefit functions are linear, then excess momentum cannot occur. Excess inertia, in contrast, remains possible even with linear benefit functions; however, if transition is Pareto optimal, then it occurs.

Proposition 5.6: Let $C \geq C^H$. a) *tech*1 might prevail over *tech*0 even if utilitarian social welfare is higher when the population sticks with *tech*0 (excess momentum). b) The population might get stuck with *tech*0 even if transition to *tech*1 produces higher utilitarian social welfare (excess inertia).

Proof: To prove the proposition it suffices to give an example for each scenario.

a) Example for excess momentum

To give an extreme example assume:

$T = 10, r=0.1, a^0 = 2, I = 10$.

$$a_0(x_0) = \begin{cases} a_0^0 + 0.8 & , \text{ for } x = 0 \\ a_0^0 & , \text{ for } 1 \geq x > 0 \end{cases}$$

$$a_1(x_1) = \begin{cases} a_1^0 + 1, & \text{ for } 1 \geq x > 0 \\ a_1^0 & , \text{ for } x = 0 \end{cases}$$

These produce:

$$a_0(T - t) = \begin{cases} a_0^0 + 0.8 & , \text{ for } t = 0 \\ a_0^0 & , \text{ for } T \geq t > 0 \end{cases}$$

$$a_1(t) = \begin{cases} a_1^0 & , \text{ for } t = 0 \\ a_1^0 + 1, & \text{ for } T \geq t > 0 \end{cases}$$

This insures that net benefits are non-negative and that transition takes a path according to NE^{seq}. Conditions BW^H and PR^H clearly hold, because *tech*1's network benefits are at its maximum immediately after agent 0 switches.

Since no agent changes his investment schedule we need to compare only the gross average benefits. Plugging in the values from (E25) and (E26) yields:

$U^{trans} = 26.32$

$U^{stay} = 28$

Hence, $W = -1.68 < 0$. Since conditions BW^H and PR^H hold, this implies excess momentum.

b) Example for excess inertia

Assume linear transition benefit functions of the form introduced in subsection 5.3. These produce (non-linear) present value transition benefit functions:

$a_0(T-t)e^{-rt} = (a_0-(a_0/T)t)e^{-rt}, t=[0,T]$

$a_1(t)e^{-rt} = (a_1/T)te^{-rt}, t=[0,T]$

Set $a_1=3, a_0=2, T=10, r=0.1$. This yields:

$W = 1.14$

$u^{trans}(0) - u^{stay} = -1.04$ (PR^H fails).

Hence, the population stays with *tech*0 although transition yields higher social welfare.   □

On the one hand, *tech*1 is superior to *tech*0 and thus increases welfare in the future. On the other hand, the transition to *tech*1 requires that the population incurs a costly period of in-compatibility. The population suffers from excess momentum if agent 0 prefers to switch im-mediately rather than to jump on the bandwagon (BW^H holds), *and* he favors transition (PR^H holds), while the society would be better off when staying with the established technology (SW^H fails). Thus, excess momentum requires that the transition costs born by early agents are lower than by late ones. Excess inertia, on the other hand, occurs if BW^H, or PR^H or both jointly fail while the population in average is better off with transition (SW^H holds).

To see why these inefficiencies can occur, we shall compare the transition benefits of the agents with respect to their position. With strong commitments (E21) becomes

$$(E21)' \qquad u^{trans}(i) = \int_0^i a_0(T-t)e^{-rt}dt + \int_i^T a_1(t)e^{-rt}dt + \frac{a_1(T)e^{-rT}}{r}$$

As already noted above, after the transition period (from $t=T$ onwards), the benefits are equal for each agent. In addition, u^{stay} (the benefits when the population stays with *tech*0) is equal for any agent. Thus, for comparison of the agents' gains and losses trough transition we need to look at benefits during the transition period. These are given by the first two terms in (E21)' and are equivalent to $u^T(i)$ (see (E24)) for $C>C^H$. Thus,

$$(E21)'' \qquad\qquad u^{trans}(i) = u^T(i) + \frac{a_1(T)e^{-rT}}{r}$$

We know that $u^{trans}(i)$ must first increase and then decrease. This follows from Assumption (5.B), which states that, at the beginning of transition, $tech0$'s benefits are larger than $tech1$'s. This implies that agent 0 has some successors whose transition surpluses are larger than his. Assumption (5.A) states that $tech1$ is superior to $tech0$, which implies that agent T has some predecessors that are made better off through transition than him. Hence, $u^{trans}(i)$ must first increase and later decrease. See Figure 34 for an example.

Figure 34

Distribution of agents' benefits from transition

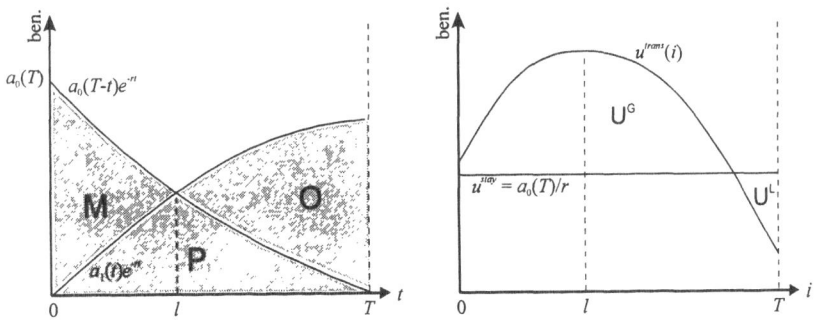

Figure 34's *lhs* is based on Figure 25. It shows the (present value) transition benefit functions during the transition period. Agent 0's first round transition benefits, $u^T(0)$, are given through the areas P+O. $u^T(T)$ corresponds to M+P, and $u^T(l)$, the benefits of the luckiest agent, equals M+P+O. The curve $u^{trans}(i) = u^T(i) + a_1(T)e^{-rt}/r$ in the graph on the *rhs* shows agents' transition benefits depending on their position. $u^T(i)$ derives from the respective areas under the curves of the *lhs*'s picture. The curve $u^{stay} = a_0(T)/r$ depicts agents' benefits if the population stays with the established $tech0$.

Figure 34 represents a situation where transition occurs. First, $w(0) = u^{trans}(0) - u^{stay} > 0$, *i.e.* PR^H holds. Condition BW^H is satisfied, too. In Figure 34's *lhs*, this is indicated by P+O > M+P, which is equivalent to $u^{trans}(0) > u^{trans}(T)$ as shown on the *rhs* of the figure.

Transition is desirable if the average surplus of transition is positive ($W > 0$), *i.e.* if the cumulated winning agents' transition surpluses are larger than the losing agents losses. These are represented through the areas U^G and U^L, respectively. Since in Figure 34's *rhs*, $U^G > U^L$, it represents a situation where transition is desirable.

Excess momentum

Contrary to the most recent literature, excess momentum is quite unproblematic in our framework. We first show in *Proposition 5.7* that excess momentum is actually even impossible if network benefit functions $(a_i(x))$ are linear. Afterwards, we demonstrate which kinds of benefit functions would support excess momentum and argue against the plausibility of these benefit functions.

Proposition 5.7: Let $C \geq C^H$. If network benefit functions are linear excess momentum cannot occur.

Proof: See Appendix A   □

Proposition 5.7 says that linear network benefit functions imply that excess momentum is impossible. Note that the proposition is based on *linear network benefit functions*, $a_0(x)$ and $a_1(x)$; thus, as long as $r>0$, the *transition benefit functions*, $a_0(T-t)e^{-rt}$ and $a_1(t)e^{-rt}$, are *nonlinear*.

How, alternatively, must the benefit functions be shaped in order to produce excess momentum? Transition requires that conditions BW^H *and* PR^H be satisfied. Any parameter whose increase makes PR^H more likely to hold makes transition more desirable. Thus, the most favorable precondition for excess momentum is when PR^H is *just* satisfied. Excess momentum requires that area $U^G - U^L < 0$. This is only possible if $a_1(t)e^{-rt}$ and $a_0(T-t)e^{-rt}$ are shaped such that they produce sufficient asymmetry of agents with respect to their transition benefits. To be precise, the faster $a_0(T-t)e^{-rt}$ *decreases* and the faster $a_1(t)e^{-rt}$ *increases*, the more likely and severe is excess momentum.[187] This is due to the fact that $a_0(T-t)$, for

---

[187]   Note for clarification of terminology: we often compare effects on the "likelihood" with the "desirability" of transition. Whether or not transition occurs depends on conditions BW^i and PR^i. These either hold or not. For whether transition occurs or not it does not matter by what extend the conditions are satisfied. Desirability, in contrast, is indicated through W. Of course, we are interested in "how much" desirable transition is. For example imagine we analyze the effects of the transition benefit function of *tech*1 shifting upwards. This makes transition more desirable. But what about condition BW^i and PR^i? Whether there is an effect depends on the actual parameter values. Maybe both condition have been satisfied anyway or perhaps they have not been satisfied before and still do not do so. Therefore, we say a change of a particular set of parameters make condition BW^i or PR^i "more likely" or "less likely" satisfied. Thus, if the transition benefit function of *tech*1 shifts upwards, the likelihood and desirability of transition increase. To put if differently, by an increase of the likelihood of transition we mean that the subset of actual parameter values (others than *tech*1's benefit function) which triggers transition is smaller.

$t = (0,T]$, does not enter condition PR^H (*i.e.* $u^{trans}(0)$) whereas it does affect the desirability of transition (since it enters $u^{trans}(i)$, $\forall i \neq 0$). A steep increase of $a_1(t)$ further intensifies the agents asymmetry, because agent 0's benefits increase over-proportionally. It, however, is also in favor of the social desirability of transition.

Figure 35 offers an example of excess momentum. The only difference to Figure 34 pertains to the transition benefit function of *tech0*. First, it decreases quickly for low values of t (*i.e.* in the beginning of transition). Second, $a_0(T)$ is now so large that condition PR^H is just satisfied. This yields Figure 35's *rhs* where area U^L has become larger than U^G.

Figure 35

Excess momentum

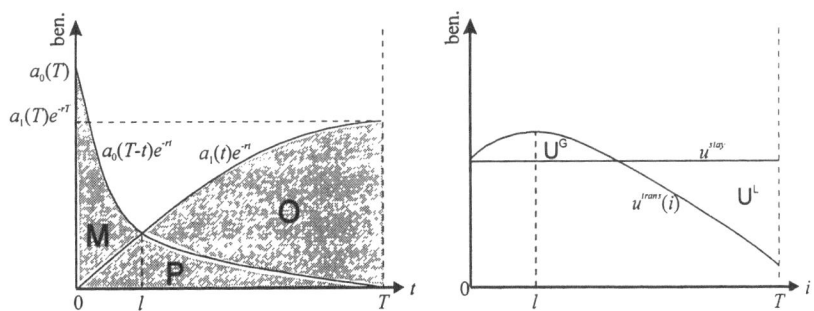

Although it remains an empirical matter, *Proposition 5.7* together with Figure 35 indicate that excess momentum should not be too problematic in our framework. The shapes of benefit functions making excess momentum a severe problem appear quite unrealistic. First, our evolutionary framework in section 3 suggests linear benefit functions.[188] Second, as *e.g.* Liebowitz and Margolis (1994) argue, marginal network benefits tend to decrease rather than to increase. This argument parallels the common assumption of decreasing marginal utility.[189]

---

[188] Note however that our modeling in section 3 represents *direct* network effects. In this model, in contrast, since network effects are exogenously captured by the network benefit function, they may be either direct or indirect. Nonetheless, most papers do assume linear benefit functions for both technologies.

[189] For example the more people possess a DVD device, the more video shops shall rent DVDs. Thus, the expected distance from the nearest DVD video shop to one's home should decrease if more people have a DVD device. It seems a reasonable assumption that marginal benefits from reduced distance reduce the closer video shops are to one's home. In the words of Liebowitz and Margolis: "... the marginal benefits of increasing the number of households that own our kind of VCR are likely exhausted now that businesses that rent videotapes are about as prevalent as ones that sell milk" (Liebowitz and Margolis 1994, p. 140).

Third, remember that *tech0* and *tech1* are substitute technologies. Why should they have asymmetric network benefit functions in the sense that *tech0*'s benefit function slowly increases for small values while *tech1*'s benefit function increases quickly?

Excess momentum in our framework is less problematic than in most established models of technological change under network effects (see our discussion in section 5.2). One reason is the often-used simplifying assumption that agents' adoption decisions are *perfectly* irreversible. This has two effects. First, agents lack the strategy to jump on the bandwagon, *i.e.* condition BW^H need not hold for transition. Second, with irreversible adoption, agents that have once adopted *tech0* do not benefit at all from superiority of *tech1*. In contrast, in our model, even those agents who are committed to the old technology for a long period can at least later enjoy the superior benefits of the new technology. Another reason is that excess momentum in our multiple-player-setup is less likely than in the often-used 2-player- (or 2 generations-) 2-periods models, since most benefits from adoption accrue to "interior" agents. This implies that (*if* transition occurs) there are always some strong winners who could provide some compensation of possible losers.

Excess inertia

Even with strong commitments, excess inertia is more problematic than excess momentum. In the proof of *Proposition 5.6b*, it has already been shown that the population might inefficiently get stuck with the established *tech0* even in the specific case of linear benefit functions.

Figure 36
Excess Inertia with strong commitments (PR^H critical)

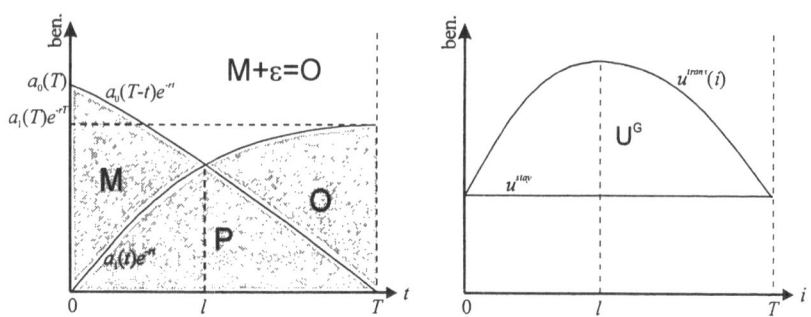

Excess inertia requires that PR^H or BW^H or both fail while SW^H holds. If PR^H is the (only) critical condition, the most severe case is when condition BW^H is just satisfied, as shown in Figure 36. Since BW^H is just satisfied, the population gets stuck with the established technology even though nearly all agents favor transition.

Things may get even worse if BW^H is the critical condition. In this case even early agents would benefit from transition and late agents even more. However, (since BW^H is not satisfied) early agents rather jump on the bandwagon, which inhibits transition to occur in the first place. Thus, transition fails although it would even be a Pareto improvement. *Proposition 5.8* shows, however, that this is impossible if benefit functions are linear.

Proposition 5.8: If network benefit functions are linear, transition always occurs if it is a Pareto improvement.

Proof: See Appendix B. □

In the proof, it is demonstrated that linear network benefit functions imply that condition BW^H always holds if condition PR^H does. This implies that if agent 0 has an incentive to pursue a jump-on-the-bandwagon strategy, he anyway favors that the population stays with *tech0*. Furthermore, since BW^H is equivalent to $u^{trans}(0) \geq u^{trans}(T)$, this implies that if agent T favors transition, agent 0 does so as well. Hence, transition occurs if every agent favors transition.

Figure 37
Excess Inertia in the very strong form (BW^H* critical)*

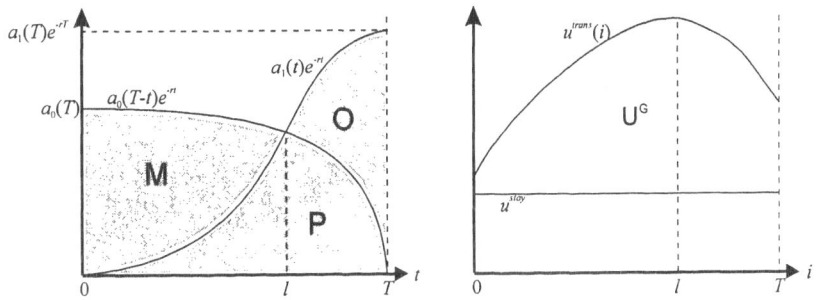

Nevertheless, excess inertia in this "very strong" form is possible if the benefit functions are shaped appropriately. Such a situation is most prevalent if the benefits with $tech0$ *decrease* very slowly at the beginning of the transition and the benefits with $tech1$ *increase* slowly.

See Figure 37 for an example. Analogous to above, this most detrimental case of excess inertia is produced by asymmetrically shaped benefit functions. As argued above, in view of the fact that the technologies are substitutes, such asymmetric shapes seem to be rather atypical. Thus, one may conclude that transition is likely to occur if every agent favors transition.

To summarize this subsection, there are external effects involved and, as a confirmation of a well-established result, these might in fact trigger market failure. However, with strong commitments, there are inherent forces that balance out the external effects. Early agents' willingness to switch determines whether transition occurs or not. While interior agents tend to gain more from transition than early ones, late agents tend to gain less than early ones. Hence, although the decision problem of those who decide for the entire population coincides with the socially optimal decision problem only by accident, they tend not to be too different from each other. In order to confirm this result, we have shown that linear network benefit functions imply that excess momentum cannot occur at all and excess inertia never occurs if transition constitutes a Pareto improvement. Furthermore, we argued that benefit functions that would produce excess momentum and very strong excess inertia should not be too common.

As we shall see in the next section, things become more unfavorable if commitments are lower.

5.5.2 Weak Commitments: Excess Momentum and Excess Inertia

As long as they remain strong (as assumed in the previous section), variations of agents' commitments neither have an effect on whether transition occurs nor on the desirability of transition. The transition path remains unchanged and agents do not incur additional cost of capital. As shown in section 5.4.3, things change if commitments fall below the threshold C^H (as assumed now). First, such weak commitments imply discontinuous transition paths including a reduction of the transition time. Second, agents that are positioned behind $t=j$ incur additional costs of capital, as they advance their reinvestment. Given C falls below the C^H, do the risks of excess momentum and excess inertia increase or decrease?

We show first that when C becomes lower, each agent's transition surplus is enlarged. Second, it is demonstrated that the average surplus of transition increases over-proportionally to agent 0's. These two results indicate that, as long as agent 0 prefers to switch first rather than to jump on the bandwagon (*i.e.* BW^L remains satisfied), transition to $tech1$ becomes both more likely and more desirable although the risk of excess inertia increases. Third, it is shown that agent 0's incentive to jump on the bandwagon increases with decreasing commitments. No matter how superior *tech1* to *tech0* is, if commitments fall below some threshold, agent 0 never switches first, implying that transition fails.

Proposition 5.9: Let $C<C^H$. If C decreases, *each* agent's transition surplus ($w(i)$) increases.

Sketch of proof: If the population stays with *tech0*, variations of C leave the benefits unchanged. Thus, we need to check only the agents' transition benefits.

Variations of C affect only first round benefits. Reproducing (E24), the first round transition benefits of agent i are:

$$(E24) \qquad u^T(i) = \begin{cases} \int_0^i a_0(T-t)e^{-rt} + \int_i^j a_1(t)e^{-rt}dt + \int_j^T a_1(T)e^{-rt}\,dt & , for\ i < j \\[2mm] \int_0^j a_0(T-t)e^{-rt} + \int_j^T a_1(T)e^{-rt}dt - \int_j^i C_i e^{-rt}\,dt & , for\ i \geq j \end{cases}$$

Let $C<C^H$ reduce from some arbitrary C' to C'' implying that $t=j$ shifts backwards from j' to j''. To prove the proposition we divide agents into several groups and evaluate how their transition benefits change:

· Agents $\in [0,j'')$

Those agents' benefits change only between the interval $t=[j'',j']$. Precisely, they receive additional benefits of

$$u^T(i)'' - u^T(i)' = \int_{j''}^{j'} (a_1(T) - a_1(t))e^{-rt} dt \qquad , for\ i < j''$$

which is clearly positive.

· Agents $\in (j',T]$

Those agents' benefits change according to

$$u^T(i)'' - u^T(i)' = \int_{j''}^{j'} (a_1(T) - a_0(T-t) - C'')e^{-rt} dt + \int_{j'}^i (C' - C'')e^{-rt} dt \qquad , for\ j' < i \leq T$$

Following from (DVI) (the definition of $t=j$) the first term is positive. Evidently, the second term is positive, too.

· Agents $\in [j'',j']$

Such an agent's benefits increase by

$$u^T(i)'' - u^T(i)' = \int_{j''}^{j'} \left(a_1(T) - a_0(T-t)\right)e^{-rt}\,dt - \int_{j''}^{i} C''e^{-rt}\,dt \quad , for\ j'' < i \leq j'$$

Since the first term in the previous expression is positive, this expression is also positive. Thus, we have shown that *each* agent's transition surplus increases if C decreases.

$\square$

If commitments become lower, the transition costs of each agent reduce, because the costly transition period is shorter. Even late agents' transition costs shrink, even though they incur additional costs of capital. This is so, because at the j-date where the simultaneous switch occurs, *tech*1 's excess benefits overcompensate the additional capital costs (otherwise agents $i > j$ would not switch in $t=j$).

We know now that weaker commitments imply that each agent's transition surplus increases. *Proposition* 5.10 shows, however, that the desirability of transition increases over-proportionally to agent 0's transition surplus.

Proposition 5.10: Let $C < C^H$. If C decreases, the average transition surplus increases over-proportionally to agent 0's transition surplus.

Proof: Let, again, C reduce from some arbitrary C' to C'' implying that $t=j$ shifts backwards from j' to j''. It will be shown that, through reduction of C, each agent's surplus increases at least as much as agent 0's does and some agents' surpluses increase more than agent 0's.

· Agents $\in [0,j'')$

Since those agents' benefits change only between j'' and j', they benefit from the reduction of C exactly as much as agent 0 does.

· Agents $\in (j',T]$

Using the expressions derived in the proof of *Proposition* 5.9, such an agent benefits *more* than agent 0 if

$$\int_{j''}^{j'}\left(a_1(T) - a_0(T-t) - C''\right)e^{-rt}\,dt + \int_{j'}^{i}(C'-C'')e^{-rt}\,dt > \int_{j''}^{j'}\left(a_1(T) - a_1(t)\right)e^{-rt}\,dt$$

which is equivalent to

$$\int_{j''}^{j'}\left(a_1(t) - a_0(T-t)\right)e^{-rt}\,dt + \int_{j'}^{i}(C'-C'')e^{-rt}\,dt > \int_{j''}^{j'}C''e^{-rt}\,dt$$

According to definition (DVI) at $t=j''$, $a_1(t) - a_0(T-t) = C''$. Thus, this inequality holds.

Agents $\in [j'',j']$

Analogous to the former group, these agents benefit more from the reduction of C than agent 0 if

$$\int_{j''}^{j'}\big(a_1(T)-a_0(T-t)\big)e^{-rt}\,dt - \int_{j''}^{i}C''e^{-rt}\,dt > \int_{j''}^{j'}\big(a_1(T)-a_1(t)\big)e^{-rt}\,dt$$

which is equivalent to

$$\int_{j''}^{j'}\big(a_1(t)-a_0(T-t)\big)e^{-rt}\,dt > \int_{j''}^{i}C''e^{-rt}\,dt$$

and must be satisfied. □

See Figure 38 (building up on Figure 30) for intuition. The figure shows the transition benefits with *tech*0 and *tech*1 for two different values of C, *i.e.* C' and C'', which in turn produce two different j-dates, j' and j''. At the respective j-dates, *tech*1 possesses a full network, *i.e.* *tech*1 exhibits benefits of $a_1(T)$ while benefits with *tech*0 become 0.[191] Through the reduction of C from C' to C'', agent 0 receives additional transition benefits during $t=[j'',j']$ as indicated by the area A, because his benefits jump from $a_1(t)$ up to $a_1(T)$ already in $t=j$'' instead of in $t=j$'. The benefits of those located behind j'', however, jump from $a_0(T-t)$ to $a_1(T)$. The area A+B+C+D represents the associated extra benefits pertaining to agents located behind j'.[192] They have to incur additional costs of capital, though. At date j'' these equal exactly the difference between the curve $a_1(t)$ and $a_0(T-t)$. Thus, immediately after $t=j$'', their additional net benefits exceed those of agent 0. Hence, in the figure, their additional net benefits during $t=[j'',j']$ are represented through A+B+D. These exceed agent 0's by B+D. Moreover, since C reduces, these agents save costs of capital after $t=j$'. For agent T, the areas E+F represent these gains. Thus, agent T's additional transition benefits through reduction of C exceed agent 0's by the area B+D+E+F. The additional benefits of agents $\in [j'',T)$ are less than this area but greater than agent 0's.

---

[191] Remember that we ignore the stand-alone values, as they do not affect any result as long as they are equal and large enough.

[192] Agents i $\in [j'',j']$ receive these extra benefits only during $[j'',i]$.

Figure 38

Commitments and agents' transition benefits

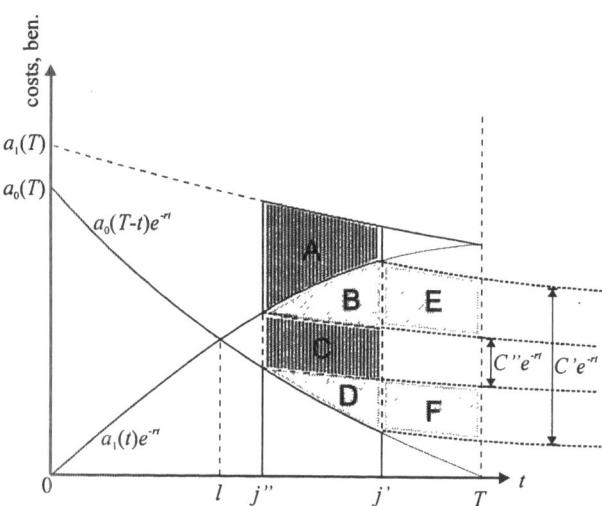

Excess momentum

Following from *Proposition* 5.9 and *Proposition* 5.10, weaker commitments make excess momentum even less problematic.[193] Since late agents' transition benefits increase more than agent 0's, the likelihood and social harm of excess momentum even further reduces if commitments decrease. (This, of course, implies that excess momentum remains impossible if network benefits functions are linear.)

Excess inertia

In contrast, excess inertia becomes more problematic. Remember that excess inertia occurs if condition PR^L, BW^L or both fail although transition is desirable (SW^L holds). Condition PR^L states that agent 0's transition surplus has to be positive. Thus, *if* BW^L remains satisfied, a reduction of the commitments makes transition both more desirable and more likely to occur. However, since late agents benefits increase over-proportionally, the risk of excess inertia is exacerbated.

---

[193]   Nevertheless, the event of excess momentum remains still not impossible for general benefit functions.

What about condition BW^L, *i.e.* agent 0's incentives to pursue a jump-on-the-bandwagon strategy?

Proposition 5.11: Let $C<C^H$. If agents' commitments decrease, condition BW^L is less likely to be satisfied.

Proof: Condition BW^L is satisfied if $u^{trans}(0) \geq u^{trans}(T)$. Since *Proposition* 5.10b states that agent *T*'s transition benefits increase more than agent 0's, the proposition is already proven. □

For intuition, recall that, when expecting transition, agent 0, at $t=0$, compares two options. He can adopt *tech*1 immediately or switch to *tech*1 at $t=j$.[194] As demonstrated above, a decrease in C shortens the costly transition period. One might think that this favors the option to adopt *tech*1 immediately, because the first round benefits with *tech*1 increase. However, from $t=j$ onwards, the gross benefits with *tech*1 are equal with both options. Benefits during $t=[0,j)$ are unchanged. Thus, this effect of a shortening of the transition period on *tech*1's benefits equally favors both options and thus is neutral.

The relevant effects are as follows. First, if C becomes smaller, it is cheaper to switch to *tech*1 later (to jump on the bandwagon). On the other hand, $t=j$ is earlier. This, in turn, has two effects. First, when postponing the switch, agent 0 has to finance *tech*1's asset longer. This makes agent 0 more willing to adopt *tech*1 in $t=0$. Second, however, the (gross) benefits with this option comparably reduce. Remember, during $t=[0,l)$, *tech*0's benefits are larger than *tech*1's. Between $t=[j,T)$, the benefits are equal with both options. Only within the interval $t=(l,j)$ are agent 0's gross benefits larger if he switches to *tech*1 immediately. Since a reduction of C shortens this interval, the gross benefits with this option reduce.

A comparison of the last two effects already yields that a reduction of C increases agent 0's incentive to switch only in $t=j$. Let C reduce from C' to C''. To compare the last two effects, we have to look at the interval $t = [j',j'']$ and compare the excess benefits of *tech*1 over *tech*0 with the additional costs of capital within this interval. If C' reduces to C'', agent 0 has to incur additional costs of

$$\int_{j'}^{j''} C'' e^{-rt} dt$$

---

[194] Another option that might appear worth to consider is to postpone his switch to the beginning of the next round (*i.e.* at $t=T+\varepsilon$). It is has, however, already been shown in section 5.4.3 that the option to switch in $t=j$ is favorable to this.

The sacrifice of excess benefits, however, amounts to

$$\int_{j'}^{j''} [a_1(t) - a_o(T-t)]e^{-rt} dt.$$

Since at j' (j'') the excess benefit of *tech*1 just equal C' (C''), the loss of excess benefits must exceed additional costs. Hence, if commitments decrease, condition BW^L is more likely to fail.

Figure 39

Commitments and agent 0's incentive to jump on the bandwagon

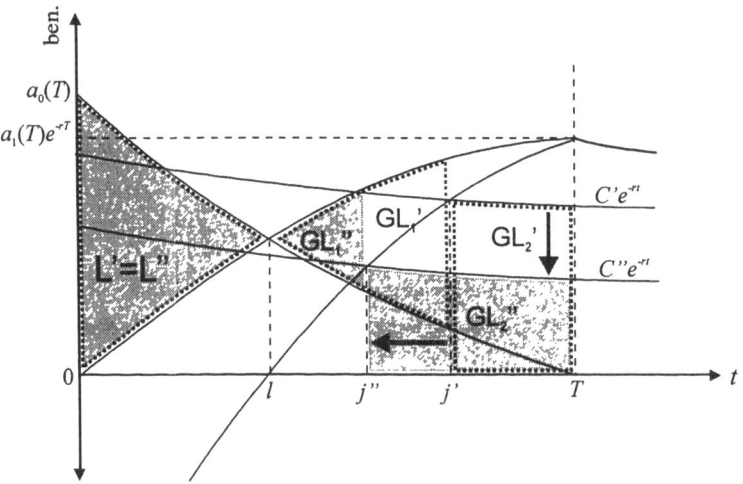

Figure 39 gives a graphical representation of this result.[195] Remember, BW^L is satisfied if the area $L < GL_1 + GL_2$. The shaded areas pertain to the level C'' and the ones surrounded by dots belong to C'. If C reduces from C' to C'', area L remains unchanged. GL_2 becomes lower as indicated by the vertical arrow. As a second effect, GL_2 expands on its left side. Third, GL_1 shrinks from GL_1' to GL_1''. Since the right side of the triangle-like GL_1 must (in equilibrium) equal the left side of GL_2 with each level of C, GL_1 reduces more than GL_2 expands on its left side. Thus, condition BW^L becomes more likely to fail.

If C approaches zero, $t=j$ approaches $t=l$. Area $GL_1 + GL_2$ becomes zero while area L remains unchanged. Thus, BW^L must fail – no matter how much superior *tech*1 is.

Proposition 5.12: Independently from the actual specification of the benefit functions, transition might fail even if transition is Pareto-optimal.

Sketch of proof: Formally, with $I:=I_1=I_0$, condition BW^L becomes

$$BW^{L'} \qquad \int_0^j a_1(t)e^{-rt}\,dt - I\frac{e^{-rT}-e^{-rj}}{1-e^{-rT}} \geq \int_0^j a_0(T-t)e^{-rt}\,dt$$

Or after including (DIV)

$$BW^{L''} \qquad \int_0^j a_1(t)e^{-rt}\,dt + \int_j^T Ce^{-rt}\,dt \geq \int_0^j a_0(T-t)e^{-rt}\,dt$$

If C approaches zero, the second term approaches zero, too. The j-date approaches $t=l$, which is located where the benefits of both technologies become equal. Hence, as long as *tech0*'s benefits with a full network are larger than *tech1*'s with zero network (as specified in Assumption (5.B)), the *lhs* of $BW^{L''}$ is always smaller than its *rhs*. This must prove the proposition because it is clearly possible to set parameters such that each agent favors transition.
□

The result that $C \to 0$ implies a failure of transition is not surprising. If there are no sunk costs associated with technology choice *and* expected transition time is greater than 0, each agent's optimal choice is to always adopt that technology which *currently* exhibits the most benefits. Even foresighted agents behave like the myopic agents from the evolutionary framework of section 3. In our framework, new technologies can never prevail over established technologies then – irrespective of how much the agents prefers transition.

5.5.3 Summary of Excess Momentum and Excess Inertia

Figure 40 summarizes the results concerning excess momentum and excess inertia for the specific case of linear network benefit functions. The horizontal axis gives agents' commitments.[195] The vertical axis measures the superiority of *tech1* over *tech0* by the difference of a_1 and a_0. The curve $w(0) = w(T)$ represents commitment-superiority-values where condition BW^i is just satisfied. $w(0)=0$ combines values where condition PR^i is just satisfied. Transition to *tech1* only occurs for C-a_1-values that are located above *both* curves. Desirability of transi-

---

[195] Figure 40 reproduces Figure 21 from the introduction in section 5.1.2.

tion is given for values that lie above $W=0$. The area under $W=0$ and the area above the upper one of $w(0)=w(T)$ and $w(0)=0$ indicates commitment-superiority combinations for which the market performs optimally. In the former, the population desirably stays with the established *tech0*. In the latter one, transition to *tech1* is desirable and does occur.

In the shaded area in between, transition to *tech1* fails although it would be desirable that it occurs. As shown in section 5.5.1, for $C \geq C^H$, excess momentum cannot occur while there is some propensity for excess inertia. Remember that (with $C \geq C^H$) excess inertia is caused by the failure of condition PR^H (because the average transition benefits typically exceed those of agent 0, see proof of *Proposition* 5.8). This produces the small shaded area for $C \geq C^H$.

Figure 40

Agents' commitments and the desirability of transition

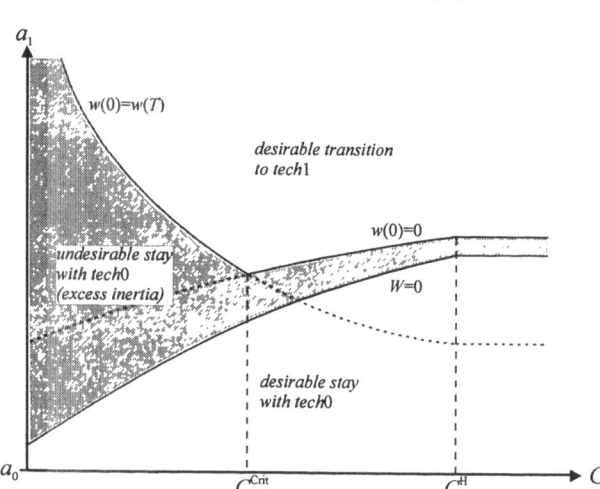

When commitments fall below C^H, transition, initially, becomes both more likely and more desirable. However, the risk of excess inertia increases, because the average surplus of transition increases over-proportionally to the transition surplus of agent 0.[196]

The fact the that agent T's transition surplus increases even more than the average transition surplus generates the intersection of the curves $w(0)=w(T)$ and $w(0)=0$ at C^{Crit}. If C falls below

---

[196]   As shown above, even if the curves were shaped such that excess momentum could occur weaker commitments reduce risk of this event.

C^{Crit}, condition BW^L becomes critical for transition. Below this threshold, the likelihood of transition and the desirability of transition behave diametrically opposite to each other. While the desirability of transition further increases, the likelihood of transition continuously decreases, as agent 0's option to postpone his switch and jump on the bandwagon becomes more and more attractive. This produces the large shaded area for low values of C.

5.5.4 Commitments and Transition Time

Now the focus will be on the transition time. Does the market-induced adoption pattern produce an optimal transition time? Could an exogenous change of the transition path increase social welfare?

As shown in subsection 5.4.3, the transition time (τ) depends on the agents' commitments (C), the transition benefit functions $a_1(t)$ and $a_0(t)$ and the interest rate (r). If commitments are strong ($C \geq C^H$), then τ equals the assets' life, T. If commitments are weak ($C < C^H$), then the transition path is discontinuous; at $t = j < T$, all (remaining) agents switch to *tech*1. Thus τ equals j. Consequently, the smaller C, the shorter is τ.

Note that r does not influence the j-date directly. This is due to the fact that agents, when they consider advancing their switch, base their decision on the difference of the *actual* benefits of the technologies. However, there is an indirect effect of r on τ: the larger r, the larger is C, as it is more expensive to finance the assets. Thus, the larger the interest rate, the later is the date of the simultaneous switch and, thus, the longer is the transition time.

It is now demonstrated that the transition time is always too long from a societal point of view.

Proposition 5.13: a) If $C < C^H$, transition time is always too long (Pareto criterion). b) If $C \geq C^H$, the transition time is Pareto-optimal. However, the average transition benefits may be larger if the transition time is shorter.

Sketch of proof:

a) Define

(DVII) $$p := t \mid C_1 = a_1(T) - a_0(T - t)$$

A path according to NE $^{p\text{-}rap}$ yields higher net benefits than NE $^{j\text{-}rap}$ to *each* agent:

· Agents $\in [0,p]$: These agents are clearly better off with NE $^{p\text{-}rap}$. Their investment schedule does not change, so costs are equal in both paths. Their transition benefits are equal during $t = [0,p]$ and $t = [j,T]$, however they are larger during $t = [p,j]$.

· Agents $\in [j,T]$: These agents are equally well off during $t=[0,p]$ as during $t=[j,T]$. During $t=[p,j]$ they incur additional costs of capital by C. According to definition (DVII), from date p onwards $a_1(T)\text{-}C > a_0(T\text{-}t)$. Thus their net benefits are larger during this period.

· Agents $\in (p,j]$: Since $a_1(T)\text{-}C > a_0(T\text{-}t)$ these agents are clearly better off.

This proofs part a) of *Proposition 5.13*.

b) Strong commitments imply sequential transition and $\tau=T$. This path is clearly Pareto-optimal. If each agent $i \in [0,T]$ switched earlier, he and his successors would be worse off. If agent i switched later, he and his predecessors would be worse off. This implies that NE^{seq} is Pareto-optimal.

To show that utilitarian social welfare might be higher through shortening of the transition time, it suffices to give an example. Let $C=C^H$, *i.e.* $\tau=T$. This implies that for sufficiently smooth transition benefit functions, if agent T advances his switches by one marginal time unit, he sacrifices almost nothing. However, he confers a positive external effect on all his predecessors. Hence, utilitarian welfare could be improved if agent T switches earlier. □

Definition (DVII) states that the p-date is where the new technology with a *full network* minus the additional costs of capital exhibits as much benefits as the old technology does. Clearly, at that point it is beneficial for all remaining agents to switch simultaneously to *tech*1. A further advance of the date of simultaneous switch would make the remaining agents worse off. (A simultaneous switch even before $t=p$ would make these agents worse off than with NE $^{p\text{-}rap}$, however not necessarily worse off than with NE $^{j\text{-}rap}$.)

See Figure 41 for illustration. Figure 41 derives from Figure 30 by adding the curve $a_1(T)$-$a_0(T\text{-}t)$.[198] Date p is determined by the intersection of that curve with C_1. The p-date must be reached before the j- date, because at $t=j$ it pays even for a single agent to switch to *tech*1.[199] If C is sufficiently small, then $p=0$, *i.e.* an advance of the simultaneous switch to just after the introduction of the new technology would be Pareto-optimal.

The reason for the sub-optimality of the transition time lies with the disability of the agents to coordinate their switches, *i.e.* with the agents' rational apathy as described in section 5.4.3.

---

[198] For convenience we have chosen linear benefit functions and actual values instead of present values in Figure 41.

[199] As the proof of *Proposition 5.13b* indicates, the optimal date maximizing utilitarian welfare lies even before $t=p$.

Commitments potentially solve one kind of collective action problem, *i.e.* they may facilitate that transition to *tech*1 occurs in the first place. However, the other collective action problem, which keeps agents in rational apathy and inefficiently retards the transition process, nevertheless exists.

Figure 41

The Pareto-optimal date of simultaneous switch

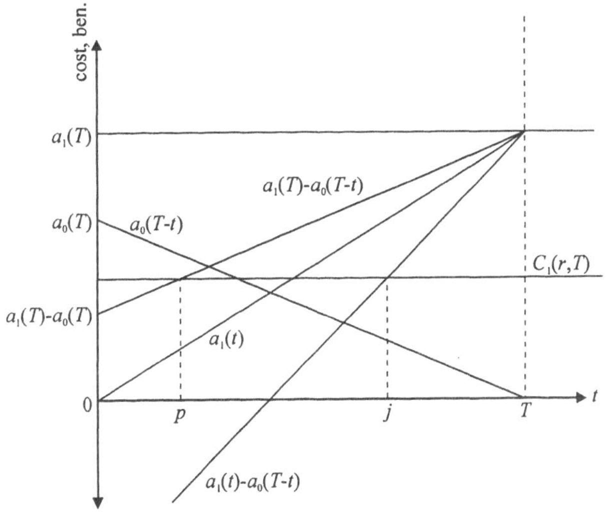

If commitments are strong ($C_1 \geq C_1^H$), then date p equals T, implying that transition time is Pareto-optimal. However, if $C_1 \geq C_1^H$ is not too large, a collective switch before T might improve utilitarian welfare. The reason for this is that agents do not take into account the positive external effect conferred on other agents when deciding on their date of switch. Clearly, once *tech*0's number of users has become small, the positive external effects of a switch on *tech*1-users are larger than the negative external effects on *tech*0-users. Thus, if C_1 is not too large, a switch by all remaining *tech*0-users to *tech*1 before $t=T$ is likely to improve social welfare.

Our results concerning the transition time are different from Farrell and Saloner (1986). As discussed in subsection 5.2.1, in addition to excess inertia in the "strong form" (transition fails although desirable) they identify excess inertia in the "weak form". In the latter, transition occurs eventually yet it starts too late. In contrast, in our framework, if transition occurs it always starts immediately after the new technology becomes available. However, the transition *process* tends to be too long.

Transition time and excess inertia

The fact that the transition time is too long is not only a welfare loss by itself. If the transition time reduced, the risk of excess inertia would reduce, too. Remember, that the transition surplus of each agent would increase. Thus, agent 0's transition surplus increases as well, making condition PR^i more likely to hold. Moreover, and probably more importantly, a reduction of the transition time makes condition BW^i more likely to hold, since agent 0's option to jump on the bandwagon become less attractive:

Proposition 5.14: If (c.p.) the transition time shortens, condition BW^i is more likely to hold.

Instead of giving a proof we explain this result with a graph.[200] See Figure 42, which builds upon Figure 41. Condition BW^L holds if $L<GL_1+GL_2$. The areas surrounded by dots belong to the path where the simultaneous switch occurs at $t=j$. The shaded areas are relevant if the simultaneous switch happens to occur already in $t=p$.

Shifting the simultaneous switch from $t=j$ to $t=p$ has two effects on condition BW^L. On the one hand, the option to postpone the switch to $t=p$ becomes more expensive as agent 0 must finance his *tech*1 assets longer. On the other hand, the period where benefits of *tech*1 exceed those of *tech*0 shortens. In $t=j$, these superior benefits equal C. Since $a_1(t)-a_0(T-t)$ increase in t while C remains constant, the first effect must be stronger than the second one.[201] Hence, it becomes more likely that condition BW^L holds. As a result, early agents' incentives to jump on the bandwagon decrease, and so a failure of transition becomes less likely.

---

[200] Since a failure of the bandwagon condition is especially relevant for low commitments, we omit the case of strong commitments.

[201] The net effect is even stronger if the p-date is before $t=z$.

Figure 42

Reduction of the transition time and condition BW^L

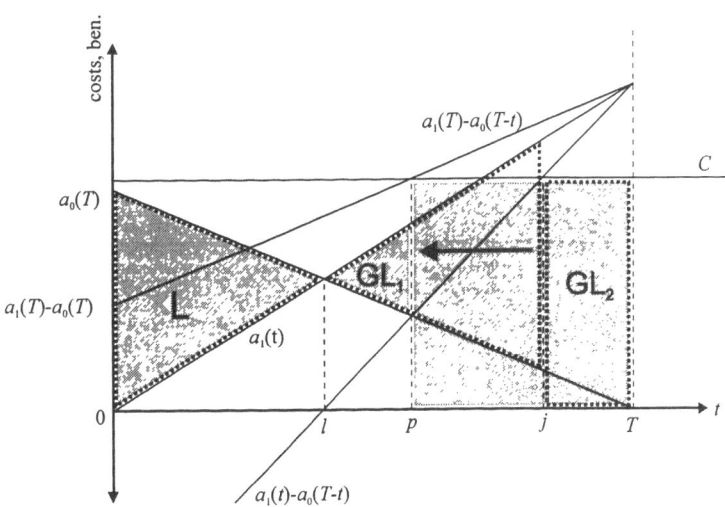

Compatibility and transition time

We now briefly demonstrate how the degree of compatibility between the technologies influences the transition time.

Proposition 5.15: The more compatible the two technologies, the longer is the transition time.

Again, we omit a proof and explain this result with a graph. Figure 43 derives from Figure 41. The bold curves belong to the "original" level of compatibility; the thin curves belong to some higher level of compatibility. As shown in the figure, an increase in the level of compatibility between the two technologies can simply be expressed by an increase in the stand-alone benefits of both technologies (leaving the technologies' benefits with a full network unchanged). C is only relevant for the network dependent part of the benefits. Thus, in the figure, the C-curve has to be shifted upwards accordingly. As a result the j-date must shift to the right (from j to j^C).

Figure 43
Compatibility and transition time

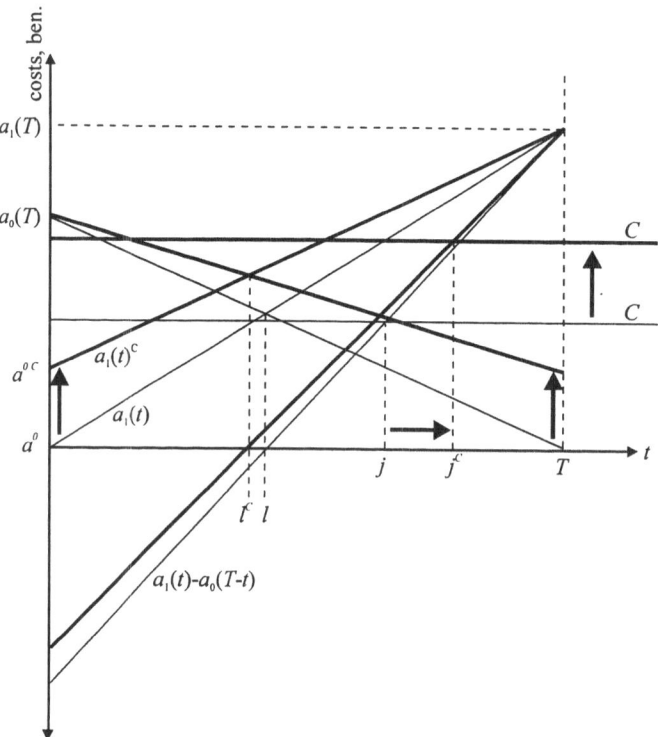

It is obvious that transition is both more likely and more desirable if the technologies are more compatible. Since any external effect reduces, each of the discussed problems diminishes, if the compatibility between the technologies increases. Nonetheless, *if* transition occurs, transition takes longer. First of all, the degree of compatibility does not affect the date where the transition starts (if it starts). However, the date where *tech*1's benefits exceed *tech*0's sufficiently to compensate agents for giving up their old (still working) asset is later. This is not surprising. Imagine the technologies are perfectly compatible, *i.e.* benefits do not at all depend on the share of agents allocated to the two technologies. When *tech*1 becomes available either any agent switches immediately to *tech*1 or at the date where his old asset breaks down. There is no reason to choose a date other than one of these.

5.6   Should Policy Intervene?

The identified potential disparities between the game's outcome and the socially optimal one pose the question whether policy-makers should get involved. In section 4, we have discussed the role of mandatory and voluntary standards in harmonization policy. Since the problem of harmonization is partly similar to the problem of technological change, the analysis partly applies to technology policy as well. One major difference, however, is that for harmonization, policy-makers can choose from a set of *established* standards, while technology policy focuses on transition to a *new* standard. According to David (1987), technology policy under network effects, therefore, faces three dilemmas, labeled "narrow windows", "blind giants", and "angry orphans". His focus is still quite different from ours, though. Instead of considering superior technologies challenging established ones, his arguments refer to the case of network technologies that compete for becoming the standard in emerging industries yet lacking a preestablished standard technology. For example, while David focuses on the Battle of BETA and VHS in mid of 1970's to establish themselves as the standard for VCRs[200], our focus is on the emerging DVD replacing the established analog (VHS) VCRs. David's arguments are largely recognized in the literature on standardization policy (see, *e.g.*, Thum 1994). Therefore, although David's focus is somewhat different from ours, we want to take the dilemmas he introduced, in order to briefly evaluate whether policy intervention in our framework is likely to be fruitful.

Narrow windows – costs and benefits of policy intervention

In section 2.3, we described that it is not at all certain that a "standard battle" will implement the best technology. If, by accident, the bandwagon has already started in the wrong direction, sales may grow quite rapidly.[201] This is problematic for policy intervention because, David argues, once the wrong technology has built up a large installed base, any attempt to turn the wheel around may be very costly. Since there is serious doubt whether policy-makers manage to intervene so early, David labels this problem the "narrow windows paradox".

Obviously, this paradox does not apply to our problem, as (the now inferior) *tech0* already dominates. In fact, quite the opposite applies. First of all, excess momentum is an unlikely event in our framework. Even if it occurs, social harm is likely to be low. Thus, the need for *immediate* intervention is likely to be low. In addition, we should expect the quality of policy-makers' information about relevant parameters (such as the quality of both technologies) to be

---

[200]   VCR = Videocassette recorder.
[201]   For example, from 1978-1981 sales of VHS VCRs doubled each year (see Gabel 1993, p. 80).

far from perfect. Thus, instead of being concerned with an early start of appropriate measures, policy-makers should first wait and see whether the new technology succeeds without any "help" from policy.

Second, it is certainly not true that costs of freeing users from an established inferior technology are always prohibitively large. What are appropriate measures when expecting excess inertia? Even if agents' commitments are strong, a strategy that policy-makers could pursue in such a situation is to declare $tech1$ mandatory in the future (which has to be announced already in $t=0$, of course). To be precise, it could make $tech1$ the mandatory standard from $t=T$ onwards. For example, in order to make digital TV prevail in the USA, the FCC has set up guidelines that schedule an end to analogue TV by 2006 and obligates TV stations to broadcast their programs digitally from 2004 on.[204] Since no agent has an incentive to not adopt the new technology, such a measure's costs are likely to be low. If commitments are strong, the benefits of such a measure are likely to be low as well, though. As shown in section 5.5.1, the stronger the agents' commitments, the smaller the risk of market failure. The lower the agents' commitments, however, the more beneficial such a measure becomes.

With low commitments, another strategy to support $tech1$ is to introduce a small subsidy that, ideally, expires at $t=p$.[205] Even though the subsidy is small this is likely to advance the simultaneous switch from $t=j$ to $t=p$. Recall, that the optimal strategy of agents i, $i\leq j$, is to stay apathetic until $t=j$. Further recall that a simultaneous switch of all remaining agents in $t=p<j$ would benefit all of them (see *Proposition* 5.13). Thus, if the subsidy expires at $t=p$, agents, instead of facing a game where it is trembling-hand-perfect to stay with $tech0$, face a simple coordination where the adoption of $tech1$ is the unique Pareto-dominant equilibrium.

Such a measure not only benefits agents directly. As shown in *Proposition* 5.14, since early agents can anticipate the earlier simultaneous switch, their willingness to switch to $tech1$ increases, which may facilitate the desirable transition to $tech1$ to occur in the first place and, thus, reduces the risk of *excess inertia* as well.[206]

Blind giants

The second dilemma that technology-policy faces refers to information deficiencies. Even if the cost-benefit balance of successful intervention is expected to be favorable, inherently badly informed policy-makers must have information about costs and benefits of a technology

---

[204] See special digital TV sites on the FCC's homepage (www.fcc.gov). The FCC (Federal Communications Commission) is a USA's governmental regulation agency.

[205] See (DVII) for the definition of $t=p$.

[206] Such a subsidy should be small in order not to produce excess momentum (if, unexpectedly, C is large).

that has never been used before. Since there are serious doubts about the quality of such information, David calls this problem the "blind giant" quandary.

If commitments are strong, this dilemma certainly applies to our framework. Policy-makers must *decide* at $t=0$ whether to make *tech*1 mandatory. If, *tech*1's quality turns out to be less superior than expected, transition costs may very well exceed the benefits. If policy-makers realize that *tech*1 is not superior in the first place, either another long and costly period of transition has to follow or the population is stuck with the inferior technology, perhaps for a long time. However, since strong commitments imply that the likelihood and social harm of market failure is limited, our analysis suggests that policy intervention is not recommendable anyway in such cases. Thus, if commitments are strong, the society should not suffer too much from this dilemma.

The risk of market failure increases when commitments become lower. Fortunately, since the social costs of transition are lower, the risk of policy intervention decreases. Moreover, for the second policy strategy we discussed, *i.e.* the acceleration of the transition process through a small subsidy, policy-makers can in fact observe the new technology before intervening. This is due to the fact, that policy-makers do not have to decide at $t=0$. They can simply wait until the transition has actually started and then induce a simultaneous switch at $t=p$.[207] Remember that it is not (mainly) the subsidy that causes early agents to switch to *tech*1 but rather the anticipation of the shorter transition process. Thus, policy-makers can very well introduce the subsidy *after* the transition has started, and hence gather information about the new technology before deciding to support it.[208]

Crucial for the success of this strategy is, however, that those agents who are supposed to switch first can sufficiently rely on policy-makers actually intervening. In other words, agents must be sufficiently sure that the transition process time *will* be accelerated, *given* that transition *has* started. Otherwise they cannot anticipate the earlier simultaneous switch and may not start the bandwagon rolling in the first place. Thus, a sound reputation of technology-policy is likely to be quite valuable.

---

[207] If C falls below some threshold, $p=0$. Nevertheless, an advance of the simultaneous switch to any $t \in [p, j]$ reduces the risk of excess inertia.

[208] The risk that policy-makers cause an inferior technology to replace the established one is excluded because a transition to such a technology cannot be a subgame-perfect equilibrium, no matter how fast the transition process would be. Moreover, as long as policy-makers do not deviate too much from $t=p$, the risk of excess momentum is small (remember that $t=p=T$ if $C \geq C^H$).

Angry Orphans

In order to obtain better information about the quality of the challenging technology, a strategy worth considering is to give the new technology a try. More precisely, policy-makers could subsidize (or otherwise support) the new technology for some time, making both technologies coexist in the market, and see how the new technology's quality turns out. Once sufficient insights are perceived, they decide either to further support transition or, in case the new technology proves not to be superior enough, to stop their support and let the population return back to the established technology. Beside initial losses of network benefits, such a strategy may, however, produce "angry orphans" (David 1987, p. 233). Those agents having switched to the new technology would suffer from being abandoned. Thus, anticipating the possibility of such decisions (still under some uncertainty) by policy-makers, agents may be very reluctant to switch in the first place. Hence, to fund such an initial period of coexisting technologies may be a risky and perhaps expensive undertaking.[209]

This problem certainly applies to our framework in cases where agents' commitments are strong. However, the lower the agents' commitments, the higher the benefits of such a strategy and the lower the risk agents face when switching to a possibly neglected technology. On the other hand, the lower the agents' commitments the more difficult it probably becomes to control such an initial period of coexisting technologies. Remember that $C<C^H$ implies that it is an equilibrium that all agents simultaneously switch between technologies. Thus, any attempt to support *tech*1 entails the risk that the entire population switches to *tech*1, making such a kind of intervention strategy a risky alternative.

As a result, the "angry orphans" problem may apply to our framework as well. In face of the availability of alternative strategies, such as the acceleration of the transition time, it is questionable, however, whether the strategy of initially maintaining both technologies is recommendable in the first place.

Summary

David's arguments against technology-policy under the presence of network effects partly apply in our framework as well. Fortunately, however, in our setting, these arguments tend to lose relevance especially in situations where desirability and efficiency of policy intervention are strong. First of all, strong commitments imply that the risk of market failure is low, making policy intervention not recommendable in the first place. If commitments are low, in contrast, the risk of market failure increases. A recommendation for policy intervention that de-

---

[209] Due to different starting points, David's "blind giants" and "angry orphans" stories differ somewhat from these told here, although the underlying arguments are unchanged.

rives from our analysis in such a situation is to introduce a very small temporary subsidy in order to quicken the transition process to an emerging technology *after* observing that the transition process has already started. Since agents anticipate faster transition, such a measure reduces the problem of excess inertia and, moreover, agents directly benefit from the shorter transition process. The costs of such a measure are low and the "blind giant" problem is likely to diminish. As it is crucial for the success of such a strategy that agents anticipate the intervention by policy-makers, a good reputation of technology-policy may be a valuable asset.

5.7 Conclusions

We have presented a dynamic model of technological change in networks. Crucial for a new network technology's success over an established one is the presence of adopters' commitments. If commitments are too low, agents – even if they expected transition to occur –jump rather on the bandwagon. Since each agent does so, the bandwagon does not start rolling in the first place. Only if agents are sufficiently committed to their once chosen technology may they have incentives to switch to an emerging technology, even if their payoffs are lower at the beginning. Commitments make agents fearful of becoming abandoned, possibly causing them to be willing to incur initial losses associated with an early switch to the new technology. This might start the bandwagon rolling.

In order to further understand how these mechanisms work, we have set up a dynamic model, which, contrary to those found in the established literature, allows for varying commitment levels. We have found that transition to the new technology occurs if those agents whose assets brake down just after the introduction of the new technology are willing to (immediately) switch to the new technology. Whether or not such agents are willing to switch, depends on two conditions. First, their payoff stream must be larger with transition than if the population stayed with the inferior established technology. Second, such agents' payoff stream is to be larger if they switch *now* instead of at any later date.

Afterwards, it has been analyzed how these conditions relate to social desirability of transition. For the special, often used, case of linear network benefit functions and equal distribution of reinvestment dates, we have found that sufficiently strong commitments imply that excess momentum cannot occur for any combination of parameter values. Even with "maximum" commitments, users may suffer from excess inertia; however, the transition to the challenging technology occurs if all agents prefer transition. If agents are moderately committed to their chosen technology, transition is faster, more likely, and more desirable. Nevertheless, the risk of excess inertia is larger because through lowering commitments, the desirability of transition increases more than the early agents' incentives to switch. For even lower commitment values, there exists some critical commitment level below which early agents' incentive to jump on the bandwagon becomes binding. Below this critical level, a further decline in commitments decreases the likelihood of transition, even though the transition time further reduces and the desirability of transition increases. No matter how superior the challenging technology is, there always exists a commitment level below which desirable transition fails.

We have further analyzed the transition *time*. Even though the transition process is shorter if commitments decrease, commitment values below the maximum imply that the transition time is always too long. One reason is that agents who are still committed to the old technology are rationally apathetic. That is, they are not able to coordinate a "premature" simultaneous switch to the new technology, even though such a collective action would make each user better off. If the transition process accelerated, not only would the social transition benefits be enhanced but also the problem of excess inertia would be reduced.

Finally, it has been asked whether policy-makers should intervene. David (1987) discusses three quandaries associated with technology policy under the presence of network effects. Although we have found that these problems may apply to our specific environment, we have argued that there are appropriate means especially in such cases where policy intervention is likely to be fruitful. Since the risk of excess momentum is limited, policy-makers could first wait and see whether the new technology succeeds without any support. In order to reduce the risk of excess inertia, we have proposed that policy-makers could accelerate the transition process to new network technologies whose diffusion process has already started. This is likely to be rewarding due to four effects. First, policy-makers can gather information about the new technology before deciding to support it. Second, the costs of such a measure are low, since the subsidy is (perhaps very) small as it is only used to help agents coordinate a premature switch. Third, it is likely that each agent benefits directly from faster transition. Fourth, more indirectly, anticipating that policy-makers will fasten the transition process once the diffusion has started, the early agents' incentives to pursue a jump-on-the-bandwagon strategy decrease, and, thus the risk of excess inertia is reduced. This effect also indicates that a sound reputation of governmental technology policy may be an essential asset.

Appendix A

Proposition 5.7: If benefit functions are linear excess momentum cannot occur.

Proof: Linear benefit functions have been introduced in section 5.3. and used in the proof of *Proposition* 5.6*b*. Due to equal distribution of agents, linear benefit functions produce (non-linear) present value transition benefit functions of the form:

$$a_0(T\text{-}t) = (a_0\text{-}(a_0/T)t)e^{-rt}, \, t\text{=}[0,T]$$
$$a_1(t) = (a_1/T)te^{-rt}, \, t\text{=}[0,T]$$

(Since $a_0^{\,0} = a_1^{\,0}$ we can ignore the stand-alone values when comparing social and private incentives to switch.)

These benefit function produce transition surplus of agent i

$$w(i) = \int_0^i (a_0 - \frac{ta_0}{T})e^{-rt}dt + \int_i^T \frac{ta_1}{T}e^{-rt}dt + \frac{a_1(T)e^{-rT}}{r} - \frac{a_0(T)}{r}$$

Average transition surplus is given by

$$W = \frac{1}{T}\left(\int_0^T w(i)di\right)$$

Transition requires that PR^{H} holds, *i.e.* $w(0) > 0$. We need to show that $W \le 0$ implies $w(0) \le 0$. Since any parameter which is in favor of $w(0)$ is also in favor of W, it suffices to show that $w(0) = 0$ implies that $W \ge 0$.

$$w(0) = 0 \quad \Leftrightarrow \quad -\frac{a_1 e^{-rt} - a_1 + a_0 rT}{r^2 T} = 0$$

Solving for a_1 yields:

$$a_1 = -\frac{a_0 rT}{e^{-rT} - 1}$$

Replacing a_1 with this expression in W yields after rearranging:

$$W' = \frac{2a_0(-e^{-2rT} + 2e^{-rT} + T^2 r^2 e^{-rT} - 1)}{r^3 T^2 (e^{-rT} - 1)}$$

We must show that W' ≥ 0. Since $e^{-rT} < 1$ the denominator is negative. Thus, we have to show that the numerator is negative or zero. Since $a_0 > 0$, this is equivalent to showing that

$$-e^{-2rT} + 2e^{-rT} + T^2 r^2 e^{-rT} - 1 \leq 0$$

Define $\alpha := rT$. Replacing, multiplication by $e^{-\alpha}$ and dividing by 2 yields

$$1 + \frac{\alpha^2}{2} - \cosh \alpha \leq 0$$

Define

$$g(\alpha) := 1 + \frac{\alpha^2}{2} - \cosh \alpha$$

As easy to see, $g(0) = 0$. Furthermore,

$$\frac{\partial g(\alpha)}{\partial \alpha} = \alpha - \sinh \alpha < 0 \quad , \forall \alpha > 0$$

Thus, $W' \geq 0$. This proves the proposition.   □

Appendix B

Proposition 5.8: If benefit functions are linear, transition occurs if it is a Pareto- improvement.

Proof: To prove the proposition it is shown that linear benefit functions imply that condition BW^H always holds if condition PR^H does. Thus, if agent 0 does not switch then he and agent T favor staying with *tech*0. This, in turn, implies that transition occurs if each agent favors transition.

Any parameter, which is in favor of PR^H is in favor of BW^H, too. Thus, it suffices to show that BW^H is always satisfied if PR^H is satisfied with equality.

With linear benefit functions:

BW^H
$$\frac{a_1(-e^{-rT}rT - e^{-rT} + 1) - a_0(e^{-rT} + rT - 1)}{r^2T} \geq 0$$

PR^H
$$\frac{a_1(-e^{-rT}rT - e^{-rT} + 1)}{r^2T} + \frac{a_1 e^{-rT}}{r} - \frac{a_0}{r} \geq 0$$

Set BW^H hold with equality. Solving for a_1 yields

$$a_1' = \frac{a_0 rT}{e^{-rT} - 1}$$

Replacing a_1 with a_1' in condition PR^H yields after rearranging

$PR^{H'}$
$$\frac{a_0(-e^{-2rT} + 2e^{-rT} + T^2 r^2 e^{-rT} - 1)}{r^2 T(e^{-rT} - 1)} \geq 0$$

To show that $PR^{H'}$ always holds is equivalent to the last part of the proof of Proposition 5.7.
□

6. Voting on Harmonization

> *"Given the fact that committees often decide on issues of standardization, even*
> *though a market based development might be preferable, the rules of decision*
> *making could have a much larger impact on welfare of society than acknowl-*
> *edged in the literature nowadays"* (Goerke and Holler 1995, p. 349).

6.1 Introduction

In the previous sections, we have identified several potential market failures associated with
network effects, and, moreover, we have asked whether and how "policy" or some "authorized body" should intervene. Considering the authorized body as an entity that strives to
achieve some well-defined goal, we however ignored the decision process within such a body.
For example, in section 4, we assumed that Europe's official standardization bodies (ESBs)
set standards in order to produce a shift from variety to harmonization. Although we argued
that bureaucratic incentives common to the ESBs' decision makers might bias that body's
objective function, we abstracted from possible conflicts *among* the decision makers *within*
the ESBs. In fact, conflicts are likely to be present within such bodies. Recall that, *e.g.*,
CEN/CENELEC's members are the "official" national standardization bodies (NSBs). We
may suppose that these national bodies, at least to some degree, act in the interest of the country they represent. Even if each country prefers harmonization, it seems that different countries often favor different standards, often their "domestic" one, for common adoption. To
resolve such conflicts, most major international standardization bodies, like CEN/CENLEC
and ISO/IEC[209], apply voting. However, is voting an appropriate mechanism for collective
decision-making in such bodies? [210]

Little attention has been devoted to both the work of standardization bodies and resolution
of conflicts within them. Farrell (1993) and Farrell and Saloner (1988) focus on the comparison of the performance of the "market mechanism" with some "committee solution". It is
claimed that the latter possesses the advantage that desirable coordination is achieved with a
larger probability as well as tending to produce a higher quality of commonly adopted standards. On the other hand, the committee solution is often slow, because it takes time to
achieve consensus among participants that have vested interests in different positions.[211] In

---

[209] ISO = International Organization for Standardization; IEC = International Electrotechnical Commission.

[210] This section builds up on Holler and Simmering (2001).

[211] David and Shurmer (1996) report that the average time taken to produce a standard varies from 2.5 years at
      the national level, through 4-5 years regionally, up to 7 years or more at the international level. According to

order to resolve conflicts in the committee, players use verbal negotiations, specifically modeled in the spirit of a "war of attrition" game. Thus, in the end, Farrell and Saloner assume that unanimity is required for decision making within the committee. In contrast, the market game is similar to the "grab-the-dollar" game where the user that adopts first will be followed by the other one. One might, therefore, think that market adoption is socially superior to the standard setting by committees if time is of great value. This conclusion, however, disregards the fact that a small discount factor also increases the propensity of the players to concede in negotiations. In fact, in the specific setting of Farrell and Saloner (1988), the committee mechanism performs unambiguously better than the market game.[212] Still, due to perhaps severe costs associated with the delay of decisions in committees, even "a rapid random choice by a bumbling bureaucrat might be better" (Farrell 1993, p. 8).

Belleflamme (2001, 2000, 1999) extends the analyses of Farrell and Saloner. He compares the adoption choice from two standards where one (immature) standard is inferior today but expected to be superior tomorrow.[213] Belleflamme shows that the committee game leads to excessive adoption of the mature standard while market adoption leads to excessive adoption of the immature one (see Belleflamme 2000, p. 22). The reason is that in the market game, users are more eager to coordinate early, while in the committee game, players, when they decide to "insist", do not take into account the costs of waiting accrued to the other player.

Although widely applied in major real world standardization, voting on standardization has been a neglected area in the literature. A first contribution is Goerke and Holler (1995) who deal with the optimal design of voting rules for harmonization. They show that the optimal design of voting rules depends strongly on the properties of the standard at stake. Moreover, even tailoring the voting rule to a particular harmonization project does not always ensure the optimal outcome.

Taking the voting rule as given, this essay investigates how majority voting performs in an authorized "harmonization body" (HB) that aims to harmonize systems goods. Since such goods are specified in more than one dimension, voting is inherently concerned with the risk of cyclical majorities. Building up on the approach of Laver and Shepsle (1990a, 1990b), we, however, argue that cyclical majorities do not necessarily arise, because the "product space" is likely to reduce to discrete grid, which implies that there may be a winning proposition that

---

CEN's annual report, to shorten the standards' elaboration time is one of the most important challenges of the CEN in the coming years (see CEN Annual Report June 2000-June 2001).

[212] See Farrell and Saloner (1988), p. 245. Farrell and Saloner model the market game similar to the "grab-the-dollar" and the committee game is modeled in the spirit of a "war of attrition".

[213] Belleflamme uses a slightly different modeling for both the committee and the market game than Farrell and Farrell and Saloner.

is chosen independently from the voting agenda. Thus, bargaining and conflicts about agenda setting are a secondary problem, which is favorable to the functioning of formal harmonization through voting.

Moreover, we analyze strategic behavior under harmonization policy. Firms, in deciding whether to make their systems' components vertically compatible, can manipulate the product space on which the HB's vote is based. Given that a benevolent HB cannot commit itself to abstain from harmonization – even if it observes that the market performs badly – the presence of harmonization policy might actually deteriorate the outcome. Thus, *ex post* optimal harmonization policy might be undesirable *ex ante*.[214]

6.2   System Goods, Vertical and Horizontal Compatibility

6.2.1 The Mix and Match setting

Our analysis builds upon "mix and match" models as established by Matutes and Regibeau (1988) and Economides (1989). Mix and match models deal with system goods like those displayed in Figure 44. Systems are composed of perfectly complementary components, such as computer systems (computer and printer), audio systems (CD-player and amplifier), telephone service systems (local and long distance telephone access), connection systems (plugs and sockets), measurement systems (inches and feet, centimeter and meter), transportation systems (trains and ships), financial service (ATM and banking service), etc.[215]

In a typical mix and match setting, consumers can either buy "pure" systems (system AA or BB) from one firm (firm A or B, respectively) or, given vertical compatibility (indicated by dotted arrows in Figure 44), hybrids, which are composed of two components supplied by different firms (systems AB and BA).[216] It is important to differentiate between vertical and horizontal compatibility. As illustrated in Figure 44, vertical compatibility refers to the interworking of functionally different components while horizontal compatibility refers to the systems themselves.[217] For example, consider computer-printer systems. Vertical compatibility is given if any type of computer can be combined with any type of printer. Horizontal compatibility is given if users of both kinds of system can exchange data with each other.

---

[214]   A similar result is obtained in Kristiansen (1998), however for different reasons.

[215]   See also section 2.1.

[216]   Note that such a setup is different from the one discussed in section 2.1. In that one, (indirect) network effects are produced by consumers' desire for variety. In contrast, in a typical mix and match setting as discussed here, consumers buy only one system. Thus, there are neither indirect network effects nor direct ones.

[217]   See Wiese (1997) for a more specific concept of compatibility.

Figure 44
Vertical and horizontal compatibility in 2x2 systems

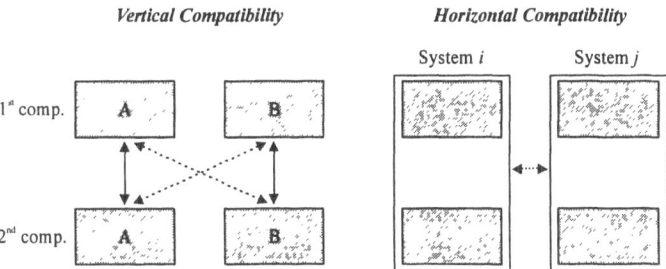

Vertical Compatibility   Horizontal Compatibility

6.2.2 Incentives to Produce Vertical Compatibility

First, as with typical mix and match settings, imagine that horizontal compatibility does not play a role. Assume that consumers' preferences are sufficiently heterogeneous, *i.e.* given equal prices, each of the possible systems (AA, AB, BA, BB) will be bought. Further assume that systems are substitutes, *i.e.* under vertical incompatibility, even (at least some of) those consumers whose ideal system is a hybrid one will still buy a (then pure) system.

Suppose marginal costs of both compatible and incompatible systems are zero. Then, the total surplus of the industry (firms' plus consumers' surplus) is very likely to be larger if systems are vertically compatible. This is obvious, as available systems (on average) better meet consumers' needs, *i.e.* there is more demand for given prices. Whether firms actually do have an incentive to produce vertical compatibility depends on whether they can appropriate a part of the rent that is generated through compatibility. Economides (1991) shows that there are two main effects that determine the firms' share.

First, due to vertical integration, there is an effect in favor of compatibility. With compatibility, each firm may face a less elastic demand function for its components (given competitor's prices). To see why, start from the same level of demand in the two regimes and assume that all systems (mixed and pure ones) face equal demand functions. Compare a price increase of, say, the first component offered by firm A that produces the same decrease of sales in either regime. Under compatibility, firm A's first component is used in systems AA and AB. Under incompatibility, this component is sold with system AA, only. Thus, firm A's incentives to increase the price for a component are higher in a regime of compatibility, because firm A does not bear the full losses of the sales (through decrease of sales of system AB a part of the sales' losses accrue to firm B), whereas under incompatibility it bears the full decrease of sales.[218] "Therefore, profits are more responsive to price under incompatibility" (Econo-mides 1996, p. 687). Matutes and Regibeau (1988) and Economides (1991,1989) show that if demand functions for all mixed and pure systems are symmetric and linear, both firms' profits are higher under compatibility. Therefore, we may presume that compatibility will result in such a specific setting.

The "vertical integration effect" is countered by the fact that competition for the individual components increases through compatibility. Remember that all four systems are substitutes. Imposing incompatibility reduces competition, as only AA and BB are available. On average, products are more horizontally differentiated in the incompatibility regime. Therefore, firms may find it profitable to grab (some of) the hybrid-preferring consumers through imposing incompatibility rather than through competing in prices. Economides (1991) shows that this counter-effect tends to dominate the vertical integration effect if demand for hybrids is suffi-ciently low relative to pure systems.[219]

Following this reasoning, in an asymmetric setting where one firm's demand for its pure system is relatively large and another firms' pure system demand is low, it can be the case that one firm wants compatibility while the other firm does not. Incompatibility is likely to result in such a situation. In many cases, firms "will not be able to counteract and correct all incompatibilities introduced by an opponent, and, therefore, in such situations of conflict we expect that incompatibility wins" (Economides 1996, p. 687).[220] Moreover, a firm that does not want compatibility might possess patents or copyrights, which might keep other firms from establishing compatibility.

Below, we analyze a setting where three firms offer two-component systems. Economides (1991) shows that the above results carry through with this case, as well.[221]

6.2.3 Horizontal Compatibility and the Benefits of Harmonization

Mix and Match models usually abstract from network effects, i.e. they assume that systems are *horizontally* compatible or that horizontal compatibility does not play a role. Conse-quently, there are no network effects and, thus, no need for harmonization. Often, however, system goods do exhibit network effects. We now describe two specific settings that are typi-

---

[218] See Economides (1989), p. 1171.

[219] See Economides (1991), p. 9.

[220] Of course, there are goods where firms can unilaterally impose some compatibility. However, compatibility is often imperfect. Especially in dynamic markets with frequent redesigns competing firms will face a great deal of problems to keep up with compatibility. Of course, if side-payments are feasible, a firm's wish as regards to compatibility might change (see Katz and Shapiro 1985, pp. 434 f.).

[221] See Economides (1991), p. 14.

cal ones in system markets. In both examples, referred to as "*case I*" and "*case II*", systems are (potentially) incompatible and (if so) subject to network effects. The analysis in section 6.4 will be based on these cases.

Case I – inherent horizontal incompatibility

As an example for *case I* goods consider a computer-printer system. If one differentiates between computers according to the operation system (UNIX, Windows, Apple) installed, it is clear that at a minimum the computer component exhibits network effects, as data exchange among computers with equal operation systems is much easier. Thus, even if each type of computer can be combined with any type of printer (*i.e.* if they are *vertically* compatible) there may very well be network effects involved (*i.e.* components are *horizontally* incompatible). Music systems demonstrate yet another example. Even if CD-players, mini-disc-players and DAT recorders could be combined with any type of amplifier, horizontal incompatibility prevails. Mini-discs cannot be played on CD-players (and vice versa) and users would benefit from being able to exchange music media.

In the specific setting in *case I*, we assume that systems are *inherently horizontally* incompatible although vertical compatibility is costless to achieve by producers. This seems to be a reasonable assumption in many cases. Vertical compatibility refers to the *interworking* of two different components, only. In contrast, production of horizontal compatibility often requires considerable redesigns of components or systems. Naturally, these are very costly and/or may make competing systems too similar to each other. This is obvious for the computer and stereo system examples.

Of course, the presence of network effects influences firms' vertical compatibility decision. In the specific setting in *case I*, it is assumed that the groups of consumers are homogeneous and each group of consumers has its "domestic" pure system which, given equal prices for each system, is favored over any other pure or hybrid system. Hence, following the reasoning in section 6.2.2, firms are likely to lack incentives to provide for vertical compatibility.

Case II – vertical incompatibility creates horizontal incompatibility

In the other setting analyzed in section 6.4, horizontal compatibility *depends* on vertical compatibility. For illustration consider again the computer-printer system. In contrast to *case I*, assume that data exchange between different computer types is perfect. However, computer users are likely to want to exchange components from time to time. In all likelihood, anyone who owns a portable computer has had annoying experiences when trying to use another printer while visiting another firm, university or simply when staying at a friend's place. Thus, in cases where components are *vertically incompatible*, systems are horizontally incompatible too, since exchange of components would be possible if every user uses the *same* system. However, *if components were vertically compatible*, systems are horizontally compatible as well. Consequently, network effects – and thus benefits from harmonization – only exist if systems are vertically incompatible.

Consider plug systems for yet another example. Often, different plug systems are vertically incompatible, *i.e.* plugs of one system do not fit into the sockets of another. However, if plugs were to perfectly fit into every socket type, everybody's benefits derived from use of plug systems would be independent from the plug systems used by others.

How are firms' incentives to produce vertical compatibility in such an industry? In contrast to *case I, case II* is based on the assumption that preferences are heterogeneous even within user groups, *i.e.* there is also a positive demand for hybrids. According to the reasoning in 6.2.2, firms may have an incentive to establish vertical compatibility. In addition to the rent generated through availability of mixed systems, vertical compatibility provides for additional surplus through the generation of network benefits. Hence, firms are even more likely to want vertical compatibility than in a setting without any network effects.

6.3   Majority Voting and Harmonization

We now focus on majority voting within authorized harmonization bodies (HBs). First, we briefly provide some details concerning the voting procedure in real world international standardization bodies. Afterwards we present a stylized voting modeling based on Laver and Shepsle (1990a, 1990b). Finally, it is argued that the outcomes of votes on technical harmonization are not necessarily unstable and path dependent – even though systems are more-dimensional goods.

6.3.1 Majority Voting in Harmonization Bodies

Belleflamme (2001), Farrell (1993) and Farrell and Saloner (1988) base their analyses on the assumption that approval of standards requires unanimity. At first glance, this seems to be a reasonable assumption, as most standardization bodies, such as CEN and ISO claim to apply the "consensus principle" for their decisions. Clause 5.1.1. of CEN/CENELEC's Internal Regulation, for example, reads as follows: "In all cases where a decision is required, every effort shall be made to reach unanimity".[222]

If a consensus cannot be achieved decisions are made by majority voting, though. For example, within the CEN and CENELEC, formal approval of a standard requires 71% of all

---

[222]   Similar rules apply to the ISO.

votes, where votes are weighted similar to those in the European Council (see Table 7).[223] Obviously, this is far from unanimity and very likely to affect the positions of players during their "efforts" incurred to reach consensus, because it considerably influences players' threat points. In fact, as long as the efforts made to reach consensus are not verifiable, rational players are likely to ignore the consensus principle entirely.

Table 7

Voting weights in the **CEN/CENELEC**

Member Country	Weighting	Member Country	Weighting
France	10	Switzerland	5
Germany	10	Austria	4
Italy	10	Sweden	4
United Kingdom	10	Denmark	3
Spain	8	Finland	3
Belgium	5	Ireland	3
Greece	5	Norway	3
Netherlands	5	Luxembourg	2
Portugal	5	Iceland	1

Source: CEN/CENELEC, Internal Regulation, part 2, clause 5.1.4

6.3.2 The Voting Model

Suppose there are three populations, A, B and C, and each population has got a representative. Further suppose that these representatives vote for the system to become the common standard. Provided a simple majority rule, which system will win the vote?

Figure 45 shows an example for a product space, which includes all candidates for harmonization. Harmonization candidates are systems composed of two components. The horizontal line represents the variants for the first component, *e.g.*, amplifiers from A, B and C, while the vertical line gives the variants for the second component, *e.g.* mini-disc (A), CD-player (B)

---

[223] See CEN/CENELEC IR2, clause 5.1.5.1. Within ISO and IEC, standard approval needs a 2/3 majority (see ISO/IEC Directives, Part 1, clause 2.7.3a). Both within ISO/IEC as well as within CEN/CENELEC, the voting procedure is slighted refined. See ISO/IEC Directives, Part 1, clause 2.7.3a and CEN/CENELEC IR2, clause 5.1.5.2, respectively, for details.

and DAT (C). AA, BB and CC stand for the "domestic" pure systems. In addition, the product space includes mixed systems such as AB, BC, CA, etc. (*i.e.* systems are vertically compatible). Note that the representatives do not have the option to vote for variety. Thus, the above product space includes all possible proposals.

Figure 45

Product space if systems are vertically compatible

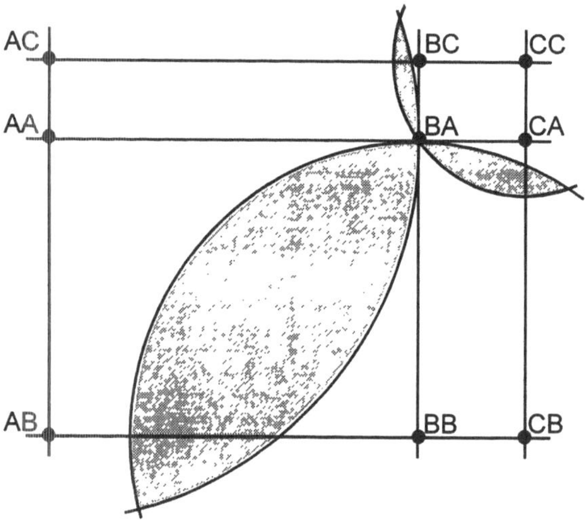

Source: Figure 45 derives from Figure 1 in Laver and Shepsle (1990a, 1990b).

The product space is constructed in such a way that each voter's preferences are represented through a set of strict circular indifference curves. Each set of indifference curves possesses the respective domestic system (*i.e.* points AA, BB or CC, respectively) as its center. Plausibly, the domestic systems correspond to the *bliss points* of the respective voters. Systems that are located close to the bliss points are preferred to systems situated farther away. For example, voter A prefers harmonization with system AA to any other alternative. Moreover, he favors system AC over AB, *i.e.* those he represents prefer to combine DAT-players rather than CD-players with type A amplifiers. Likewise, people from B would rather harmonize with CA than with AB. With the music systems example, they prefer a combination of a minidisc player with a C-amplifier to one of a CD-player with an A-amplifier.

It is assumed that each voter's preferences equal those of the median voter of the corresponding population. This may indicate that the voter is himself chosen through a vote within the population that he represents. Note that preferences *within* each population might very well be heterogeneous. That is to say, some members of population A may very well prefer another system to system AA, for example. Before harmonization, any of the nine available systems may be in use within either population. Thus, the specific example used in Figure 45 may fit the assumptions of both case I (where preferences are homogeneous within populations) and case II (where preferences are heterogeneous even within populations).

6.3.3 Are Outcomes Unstable and Path Dependent?

Social choice theory tells us that the outcome of majority voting in such a spatial setup is likely to be unstable and path dependent, if the set of alternatives is continuous in both dimensions.[224] With such a continuum, for any proposed harmonization candidate, another system could be proposed that beats the former one by simple majority. Thus outcomes are unstable and path dependent and, therefore, likely to depend on the voting agenda.

As Laver and Shepsle (1990a, 1990b) show, we can, however, hope for stable and path independent outcomes, if the set reduces to a discrete grid. In political theory, it is argued that voters' perception, or the candidates' power to form platforms, or both, reduce the policy space to such a discrete grid. It seems even more convincing to assume that system goods can be represented in a discontinuous space like that. Often, small variations of products are either not perceived by the buyer or even constitute a failure through negatively influencing the functioning of the system. Naturally, technical system goods lack the flexibility to make small variations in the quality of their components. Think of connecting trains as an example or other even less flexible interfaces in integrated traffic systems. Moreover, a products' flexibility is likely to be mitigated through learning curve effects. Thus, established system components are likely to build superior harmonization candidates. As a consequence, we can hope for stable and path independent outcomes.[225]

6.3.4 Outcome of the Voting Game

In the example in Figure 45, the product space reduces to the points that correspond to the three pure and six mixed systems. System BA wins the vote. To see why, construct the "winning sets" pertaining to each system. The winning set of a system includes all proposals that

---

[224] See Riker (1982) and Laver and Shepsle (1990a, 1990b).
[225] For a discussion of sufficient conditions for stability in such a setting see, *e.g.*, Plott (1967), Saari (1997) and, for an overview, Nurmi (1998).

win against the system (in a pair-wise challenge). Graphically, one obtains the winning set of a specific system by taking each voter's indifference curve that runs through the point representing the system. The three inner areas between each two indifference curves represent the winning set pertaining to the specific system (see also Laver and Shepsle 1990a, 1990b). In Figure 45, BA's winning set is indicated by the three shaded areas. It is easy to see this is the only empty one. Hence, system BA wins the vote.

Provided that countries are symmetric in size and preferences, we may apply the Euclidean distance as an indicator for the industry's surplus. Harmonization with system BA minimizes the sum of distances to the bliss points. Thus, in our example, the voting mechanism implements the socially best solution.

6.4   Strategic Behavior Under Harmonization Policy

In section 6.2, we have discussed firms' incentives to produce vertical compatibility in system good markets. Two cases, *case I* and *case II*, have been established in which harmonization might be beneficial. In section 6.3, we have analyzed a voting model, which has been illustrated by an example given in Figure 45. We now link this example to the specific cases introduced in section 6.2.3, in order to discuss two pitfalls that harmonization policy might face in such industries. First, we will describe the assumed way in which harmonization is conducted (section 6.4.1). Afterwards, we will argue that firms – anticipating the way harmonization policy is conducted – might have adverse incentives to manipulate the product space on which the vote for harmonization is based (sections 6.4.2 and 6.4.3).

6.4.1 Harmonization Policy

Throughout this section, we assume that harmonization policy is conducted as follows. First, the HB observes the market outcome. Based on this observation, it decides whether to harmonize systems or not. Suppose (permanent) representatives in the HB have concluded some pre-arrangement, which states that harmonization is to be implemented if they observe that the market performs sub-optimally, *i.e.* if the industry's total surplus could be improved through harmonization. If harmonization does not increase industry's surplus, all players go their way. If, in contrast, the HB has recognized that harmonization does enhance industry surplus, then there will be a vote. The winning system will be taken for harmonization.

This kind of harmonization policy is only reasonable if the following two assumptions are satisfied. First of all, the populations must have an incentive to conclude the pre-agreement for harmonization. Note that this agreement implies that in a particular harmonization project, some populations might be worse off through harmonization. Thus, such a pre-arrangement is likely to be beneficial for any population as long as the number of harmonization projects is large. Alternatively, one could assume that there is a central government, which aims at increasing utilitarian welfare and has imposed such a rule on the HB. This assumption may be consistent with actual EU harmonization policy where the European Commission may mandate the production of particular standards to the official standardization bodies (see section 4).

Second, to assume this kind of harmonization policy makes only sense if vertical compatibility is not enforceable – neither by the court nor unilaterally by one firm. Often, vertical compatibility is difficult to verify before courts. After all, how can one measure vertical compatibility? Even if interfaces physically fit into one another, interworking of components from different firms may be imperfect. Think of stereo systems, for example. There is a plethora of parameters that determine how well functionally different components work together. Second, in many cases, production of vertical compatibility requires adjustments of *both* components offered by two *different* firms. Thus, it is difficult to prove which of the firms has failed to make the necessary product redesigns. This seems especially problematic in dynamic markets, where firms frequently introduce new generations of products.

For simplification of the analytical framework do we further assume that vertical compatibility only prevails if *all three firms* want compatibility. This assumption is probably unrealistic in many industries, since, *e.g.*, firms A and B can make their systems compatible independently from firm C. It will be shown how the results change if this assumption is relaxed.

6.4.2 Case I

Assume a setting in accordance with *case I* (as described in section 6.2.3). Recall that preferences *within* each population are homogeneous, while they differ *among* the populations.[225] This implies that each voter's preferences are equivalent to the preferences of the users he represents. The most preferred system of users from population A is system AA, people from B prefer BB and system CC is the most favored one by residents of C. We assume that these preferences fit Figure 45. Further recall that all systems are inherently horizontally incompatible. That is to say, even if systems are vertically compatible, there are network effects, and, thus, social benefits from harmonization.

Social preferences

It is assumed that network effects are strong such that harmonization increases the industry's surplus irrespective of the specific system that is taken for harmonization. Thus, the social rank order is given through

---

[225] See section 3.2 for a brief discussion of such an assumption.

Assumption (6.A)

$$H^C \succ H^{Inc} \succ V^C \sim V^{Inc}$$

where H stands for harmonization and V for variety; the superscripts C and Inc indicate that systems are vertically compatible or vertically incompatible, respectively. Notice that with variety, social welfare is independent from whether systems are vertically compatible. This is due to the assumption that preferences are homogeneous within user groups, *i.e.* in a variety regime no consumer uses mixed systems. In contrast, H^C is socially superior to H^{Inc}. Vertical compatibility makes mixed system available for harmonization. As shown in section 6.3.3, such a system, BA, is a socially superior harmonization candidate.

Outcome without harmonization policy

As discussed in section 6.2.3, without harmonization policy, firms are likely to keep systems vertically incompatible. Thus, the market outcome is V^{Inc}.

Outcome under harmonization policy

Due to Assumption (6.A), any possible market outcome that the HB observes can be improved through harmonization. Hence, there will be a majority vote that determines the system, which will be taken for harmonization.

The set of available harmonization candidates depends on the vertical compatibility decisions of firms. If firms want compatibility, all nine systems are available for harmonization. As shown in section 6.3.3, system BA wins. If at least one firm opposes compatibility, then pure systems are available, only. Figure 46 shows the resulting product space. It is easy to see that system BB wins. It is very likely that firm B favors harmonization with system BB over BA. Thus, anticipating the voting outcome, firm B will keep incompatibility. Hence, the outcome under harmonization policy is H^{Inc}.

The socially optimal outcome, H^C, cannot be reached through harmonization policy in our setting. Irrespective from the specific product space, as long as the firms favor harmonization with their pure system over any other system, H^C cannot be obtained.

Notice that in our specific example, this result carries through even if we relax the assumption that vertical compatibility requires the admission of *each* firm. Figure 46 exhibits that system BB wins, even if the systems of firms A and C are compatible.

Figure 46
Product space if systems are vertically incompatible

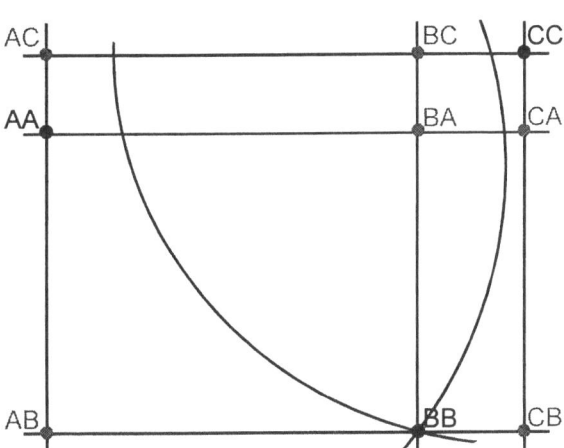

6.4.3 Case II

Now we analyze a setting in accordance with *case II* (see section 6.2.3). Recall that network effects *only* prevail if systems are *not* vertically compatible. Furthermore, recall that preferences are heterogeneous even *within* each population, *i.e.* under variety users buy mixed systems as well as pure ones. As mentioned in section 6.3.2, even now, our specific product space example can be applied. Consistently, one may assume that each population's voters' preferences correspond to the preferences of the respective median users.

Social preferences

Since network effects are exploited if systems are vertically compatible the social rank order is assumed to be

Assumption (6.B)

$$V^C \succ H^C \succ H^{Inc} \succ V^{Inc}$$

Outcome without harmonization policy

As discussed in section 6.2.3, without harmonization policy, firms are very likely to render the systems vertically compatible. Thus, the market outcome equals the socially optimal one, V^C.

Outcome under harmonization policy

Due to Assumption (6.B), harmonization only improves social welfare if systems are vertically incompatible. All the same, each firm is very likely to prefer harmonization with its pure system to any other outcome. Firms can anticipate that there will be harmonization if one of them opposes compatibility. Anticipating that its pure system wins the vote (see Figure 46), firm B keeps incompatibility. Consequently, the outcome under harmonization policy is, once again, H^{Inc}.[226]

This result is much more unfavorable than the one obtained for *case I*. Harmonization policy is not only unable to implement the socially optimal outcome. Even if we abstract from other social costs usually associated with policy intervention, the outcome would be socially superior without harmonization policy. Recall that this result is already based on the assumption that policy-makers *only* intervene if it observes that the market performs sub-optimally. Thus, our example indicates that *ex post optimal* harmonization policy might be undesirable *ex ante*.

Still, the outcome would change if the HB could ex ante commit not to impose harmonization even if it observes that the market performs badly. Therefore, one must ask whether harmonization policy really works the way it is assumed. In the end, this depends on who can endure the longest, firms or politicians (or people in the SB). While firm B loses money in "unharmonized" incompatible markets, politicians must cope with pressures from users and user associations. Thus, it probably also depends on the public's interests in the considered industry whether or not the HB can ex ante commit to abstain from harmonization.

6.5 Conclusions

Our analysis has been far from general. All the same, it illustrates some ideas that are potentially relevant for the design of decision mechanisms in standardization bodies. We have shown that voting on harmonization of systems is not necessarily associated with cyclical majorities. Systems naturally lack the flexibility with respect to any arbitrary adjustment.

---

[226] The outcome might differ if we relax assumption that admission of *each* of the three firms is necessary for vertical compatibility. In our example, firms A and C have a strong incentive to render their systems vertically compatible with each other. Parallel to *case I*, this would not change the vote's outcome. However, it might avoid (the vote on) harmonization to occur in the first place, if variety with such partial compatibility is socially superior to harmonization. Thus, our result only applies if network effects are sufficiently strong.

Therefore, the product space that includes all possible proposals in such a vote is likely to reduce to a discrete grid. Cyclical majorities and thus unstable and path dependent outcomes are less likely to evolve than in a continuous space. This result is favorable to the functioning of voting as a decision mechanism, as it is, in fact, widely used in various major standardization bodies such as CEN/CENELEC and ISO.

We have also discussed pitfalls that might arise with harmonization policy. One striking result is that ex post desirable harmonization policy might be harmful ex ante. Anticipating the voting outcome within an authorized harmonization body, firms may have incentives to manipulate the product space that the body's vote is based on. In cases where the harmonization body cannot commit to abstain from harmonization even if it has observed that the industry performs badly, the outcome with harmonization policy might be inferior to the outcome that would be achieved without harmonization policy. This result indicates that even credible regulation threats might do a bad job in such markets.

7. Summary of Findings

Three essays on the evolution of standards in networks have been presented. The first essay has investigated how increasing integration among countries ("globalization") affects the evolution of global standards. We have introduced an evolutionary game theoretic model, which allows for the explicit analysis of exogenously driven globalization. Our analysis suggests that globalization does not necessarily implement international harmonization of (perfectly) incompatible standards even if it is efficient. Moreover, even if increasing globalization eventually implements harmonization there are two pitfalls. First, harmonization typically occurs too late from a social point of view. Second, increasing globalization may produce harmonization with an inferior standard. It has also been analyzed how double adoptions or converters may affect these results. We have found that the availability of double adoptions supports harmonization, even if they do not occur in (stable) equilibrium. Nevertheless, even though efficient harmonization is more likely to occur, it still comes with the risk that market-induced harmonization makes populations end up with an inferior standard.

The second part of the essay has compared the appropriateness of voluntary and mandatory standards for producing harmonization. As the latter kind of measures is likely to be more expensive, we have argued that they should only be applied if variety's stability is high. If the stability of variety is low, voluntary standards are likely to suffice to implement harmonization. We have found that EU harmonization policy makes, in fact, use of both kinds of standards. Through granting an official and exclusive status to selected European standardization bodies, it fosters their ability to generate collective switches through their standards. Moreover, in some cases, standards are directly or indirectly sanctioned (or subsidized) by European legislations. It remains, however, unclear whether European harmonization makes excessive use of declaring standards mandatory.

We have also found that European official standardization bodies may have excessive harmonization incentives. This poses the question as to whether one should reduce their power. Unfortunately, there is no immediate solution to it. Reducing their "harmonization power" is likely to restrict excessive harmonization. However, our stability analysis has demonstrated that the inferior harmonization candidate may very well be the one that is easier to reach. Thus, a reduction of the harmonization power allocated to the standardization bodies may produce the adverse incentive of choosing too often an inferior standard for harmonization.

The second essay has dealt with technological progress in networks. We have set up a game-theoretic model in order to explicitly analyze the role of users' commitments to their adopted technology. The focus has been on industries where goods or "assets" pertaining to each technology have a limited life. Commitments are produced through sunk costs associated with each technology choice. Our study suggests that the risk of too much technological change is low in such industries. Even with very strong commitments, users may suffer from excess inertia; however, transition to the challenging technology does occur if every agent prefers transition. For commitment values that lie moderately below maximum commitments, transition is faster, more likely, and more desirable. All the same, the risk of excess inertia is higher even for such moderate commitments. For even lower commitments values, there exists some critical commitment level below which the "bandwagon condition" becomes binding. For such low commitment values, the likelihood of transition is smaller, even though the transition time would further reduce and the desirability of transition increases. No matter how superior the challenging technology is, there always exists a commitment level below which desirable transition fails.

Finally, it has been asked whether policy should intervene in such industries. There are appropriate means especially in such cases where policy intervention is likely to be fruitful. Since the risk of excess momentum is limited, policy-makers could first wait and see whether new technologies succeed without any support from policy. It is proposed that intervention should accelerate the transition process to new network technologies whose diffusion process has already started. This is likely to be rewarding due to four effects. First, it is easier for policy-makers to obtain information about an emerging technology when it has already been in use. Second, the costs of such a measure are low, since a small subsidy may suffice to help agents coordinate a desirable premature switch. Third, it is likely that a shorter transition process directly benefits each agent. Fourth, more indirectly, *anticipating* that policy *will* shorten the transition process once the diffusion has started, early agents' incentives to postpone their switch (pursue a jump-on-the-bandwagon strategy) decrease. As it is crucial for the success of such a strategy that agents anticipate the intervention by policy-makers, a sound reputation of technology-policy may be an essential asset.

The third essay has been devoted to the work of standardization bodies and the resolution of conflicts within them. We have shown that voting on harmonization of systems is not necessarily associated with cyclical majorities, even though such goods are specified in more than one dimension. This result is favorable to the functioning of voting as a decision mechanism, as it is widely used in various major standardization bodies such as CEN/CENELEC and ISO.

We have also discussed pitfalls that might arise with harmonization policy. One result is that ex post desirable harmonization policy might be harmful ex ante. Anticipating the voting outcome within a harmonization body, firms may have incentives to manipulate the product space that the body's vote is based on. In cases where the harmonization body cannot commit to abstain from harmonization even if it has observed that the industry performs sub-optimally, the outcome with harmonization policy might be inferior to the outcome that would be achieved without harmonization policy. This result indicates that even credible regulation threats might do a bad job in such markets.

Bibliography

Adams, M. (1996): "Norms, Standards, Rights." In Holler, M.J. and Thisse, J.-F. (eds.): The Economics of Standardization. Special issue of *European Journal of Political Economy*, 12, pp. 363–75.

Akerlof, G. (1970): "The Market for 'Lemons': Quality Uncertainty and the Market Mechanism." *Quarterly Journal of Economics*, 89, pp. 488-500.

An, M.Y. and Kiefer, N.M. (1995): "Local Externalities and Societal Adoption of Technologies." *Journal Evolutionary Economics*, 5, pp. 103-117.

Andreozzi, L. (2001): "Society Saved by Children: The Role of Youngsters in the Generation of Scandals." Forthcoming in *Homo Oeconomicus*.

Arthur, W.B. (1989): "Competing Technologies, Increasing Returns, and Lock-In by Historical Events." *The Economic Journal*, 99, pp. 116-31.

Azoulay, P., Berndt, E.R. and Pindyck, R.S. (2000): "Consumption Externalities and Diffusion in Pharmaceutical Markets: Anti-Ulcer Drugs." NBER Working Paper #7772.

Belleflamme, P. (1998): "Adoption of Network Technologies in Oligopolies." *International Journal of Industrial Organization*, 16, pp. 415–444.

Belleflamme, P. (1999): "Assessing the Diffusion of EDI Standards Across Business Communities." Holler, M.J. and Niskanen, E. (eds.): *EURAS Yearbook of Standardization*, 2, special issue of *Homo Oeconomicus*, pp. 301-324.

Belleflamme, P. (2000): "Coordination on Formal vs. De Facto Standards: A Dynamic Approach." Working Paper (Queen Mary and Westfield College).

Belleflamme, P. (2001): "Coordination on Formal vs. De Facto Standards: A Dynamic Approach." *European Journal of Political Economy*, forthcoming.

Berg, S.V. (1989): "The Production of Compatibility: Technical Standards as Collective Goods." *Kyklos*, 42, pp. 361-383.

Berndt, M. (2000): "Global Differences in Corporate Governance Systems: Theory and Implications for Reforms." Working Paper (Harvard Law School).

Berndt, M. and Simmering, V. (2001): "The Impact of Network Effects and Preferences on the Standardization of Institutions." Working Paper (Institute for Socio Economics, University of Hamburg).

Besen, S.M. and Farrell, J. (1994): "Choosing How to Compete: Strategies and Tactics in Standardization." *Journal of Economic Perspectives*, 8, pp. 117-131.

Blankart, C.B. and Knieps, G. (1993): "State and Standards." *Public Choice*, 77, pp. 39-52.

Bresnahan, T. and Greenstein, S. (1996): "Technical Progress and Co-invention in Computing and in the Uses of Computers." *Brookings Papers on Economic Activity: Microeconomics*, 1-78.

Brynjolfsson E. and Kemmerer, C. (1996): "Network Externalities in the Microcomputer Software: An Econometric Analysis of the Spreadsheet Market." *Management Science*, 42, pp. 1627-47.

Cabral, L.M.B. (1990): "On the Adoption of Innovations with 'Network' Externalities." *Mathematical Social Sciences*, 19, pp. 299-308.

Choi, J.P. (1994): "Irreversible Choice of Uncertain Technologies with Network Externalities." *Rand Journal of Economics*, 25, pp. 382-401.

Choi, J.P. (1996): "Standardization and Experimentation: Ex Ante vs. Ex Post Standardization." In: M.J. Holler and J.-F. Thisse (eds.): The Economics of Standardization. Special issue of *European Journal of Political Economy*, 12, pp. 273-290.

Choi, J.P. and Thum, M. (1998): "Market Structure and the Timing of Technology Adoption with Network Externalities." *European Economic Review*, 42, pp. 225-244.

Chou, C. and Shy, O. (1990): "Network Effects without Network Externalities." *International Journal of Industrial Organization*, 8, pp. 259-270.

Chou, C. and Shy, O. (1993): "Partial Compatibility and Supporting." *Economics Letters*, 41, pp. 193-197.

Chou, C. and Shy, O. (1996): "Do Consumers Gain or Lose When More People Buy the Same Brand." In: M.J. Holler and J.-F. Thisse (eds.): The Economics of Standardization. Special issue of *European Journal of Political Economy*, 12, pp. 309-330.

Church, J. and Gandal, N. (1992): "Network Effects, Software Provision, and Standardization." *Journal of Industrial Economics*, 40, pp. 85-103.

Church, J. and Gandal, N. (1993): "Complementary Network Externalities and Technological Adoption." *International Journal of Industrial Organization*, 11, pp. 239-260.

Clark, R. (2000): "Legal Aspects of Standardization in Ireland." In: Falke, J. and Schepel, H. (eds.): Legal Aspects of Standardization in the Member States of the EC and of EFTA – Volume 1: Country Reports. Luxembourg.

Cooter, R. (1998), "Expressive Law and Economics", *Journal of Legal Studies*, 27, pp. 585-607.

Crémer, J., Rey, P. and Tirole, J. (2000): "Connectivity in the Commercial Internet." *Journal of Industrial Economics*, 48, pp. 433-473.

David, P.A. (1985): "Clio and the Economics of QWERTY", *American Economic Review*, 75, pp. 332-37.

David, P.A. (1987): "Some New Standards for the Economics of Standardization in the Information Age." In: Dasguta, P. and Stoneman, P.L. (eds.): Economic Policy and Technological Performance. Cambridge.

David, P.A. and Shurmer, M. (1996): "Formal Standard-Setting for Global Telecommunications and Information Services." *Telecommunications Policy*, 20, pp. 789-815.

Davis, S. (1992): "Using Price to Communicate Information about a Firm's Capabilities and Quality Decisions." Working Paper (UC Davis).

De Bijl, P.W.J. and Goyal, S. (1995): "Technological Change in Markets with Network Externalities." *International Journal of Industrial Organization*, 13, pp. 307-325.

De Palma, A. and Leruth, L. (1996): "Variable Willingness to Pay for Network Externalities With Strategic Standardization Decisions." In Holler, M.J. and Thisse, J.-F. (eds.): The Economics of Standardization. Special issue of *European Journal of Political Economy*, pp. 235-251.

De Palma, A., Leruth, L. and Regibeau, P. (1999): "Partial Compatibility with Network Externalities and Double Purchase." *Information Economics and Policy*, 11, pp. 209-227.

Dranove, D. and Gandal, N. (2001): "The DVD vs. DIVX Standard War: Empirical Evidence of Vaporware." Working Paper (Northwestern University).

Dybvig, P.H. and Spatt, C.S. (1983): "Adoption Externalities as Public Goods." *Journal of Public Economics*, 20, pp. 231-247.

Ebert-Kern, B. (1994): Ökonomische und rechtliche Auswirkungen technischer Harmonisierungskonzepte im europäischen Normungssystem. Diss. Frankfurt/Main.

Economides, N. (1989): "Desirability of Compatibility in the Absence of Network Externalities." *American Economic Review*, 79, pp. 1165-1181.

Economides, N. (1991): "Compatibility and the Creation of Shared Networks." In: Guerrin-Calvert, M.E. and Wildman, S.S. (eds.): Electronic Services Networks. New York.

Economides, N. (1993): "Network Economics with Application to Finance." *Financial Markets, Institutions & Instruments*, 2, pp. 89-97.

Economides, N. (1996): "The Economic of Networks." *International Journal of Industrial Organization*, 14, pp. 673-699.

Economides, N. and Flyer, F. (1997): "Compatibility and Market Structure for Network Goods." Working Paper (NYU).

Economides, N. and Himmelberg, C. (1995): "Critical Mass and Network Size with Application to the US FAX Market." Working Paper (NYU).

Economides, N. and Salop, S.C. (1992): "Competition and Integration Among Complements, and Network Market Structure." *Journal of Industrial Economics*, 40, pp. 105-122.

Ellison, G. (1993): "Learning, Local Interaction, and Coordination." *Econometrica*, 61, pp. 1047-1071.

Ellison, G. and Fudenberg, D. (2000): "The Neo-Luddite's Lament: Excessive Upgrades in the Software Industry." *Rand Journal of Economics*, 31, pp. 253-72.

Falke, J. (2000): Rechtliche Aspekte der Normung in den EG-Mitgliedstaaten und der EFTA – Volume 3: Deutschland. Luxembourg.

Farrell, J and Katz, M.L. (1998): "The Effects of Antitrust and Intellectual Property Law on Compatibility and Innovation." *Antitrust Bulletin*, 43, pp 609-650.

Farrell, J. (1993): "Choosing the Rules for Formal Standardization." Working Paper (UC Berkeley).

Farrell, J. (1995): "Arguments for Weaker Intellectual Property Protection in Network Industries." In: Kahin, B. and Abbate, J. (eds.): Standards Policy for Information Infrastructure. Cambridge.

Farrell, J. and Saloner, G. (1985): "Standardization, Compatibility and Innovation." *Rand Journal of Economics*, 16, pp. 70-83.

Farrell, J. and Saloner, G. (1986a): "Installed Base and Compatibility: Innovation, Product Preannouncement, and Predation." *American Economic Review*, pp. 940-955.

Farrell, J. and Saloner, G. (1986b): "Standardization and Variety", *Economics Letters*, 20, pp. 71-74.

Farrell, J. and Saloner, G. (1987): "Competition Compatibility and Standards: The Economics of Horses, Penguins and Lemmings." In: Gabel, H.L. (ed.): Product Standardization and Competitive Strategy. Amsterdam.

Farrell, J. and Saloner, G. (1988): "Coordination through Committees and Markets." *Rand Journal of Economics*, 19, pp. 235-252.

Farrell, J. and Saloner, G. (1992): "Converters, Compatibility, and the Control of Interfaces." *Journal of Industrial Economics*, 40, pp. 9-35.

Farrell, J. and Shapiro, C. (1992): "Standard Setting in High-Definition Television." *Brookings Papers: Microeconomics*, pp. 1-93.

Foray, D. (1997): "The Dynamic Implications of Increasing Returns: Technological Change and Path Dependent Inefficiency." *International Journal of Industrial Organization*, 15, pp. 733-52.

Foster, D. and Young, H.P. (1990), "Stochastic Evolutionary Game Dynamics." *Theoretical Population Biology*, 38, pp. 219-32.

Fudenberg, D. and Tirole, J. (1991): Game Theory. Cambridge, London.

Fudenberg, D. and Tirole, J. (2000): "Pricing a Network Good to Deter Entry." *The Journal of Industrial Economics*, 48, pp. 373-391.

Gabel, H.L. (1993): Produktstandardisierung als Wettbewerbsstrategie. London, New York et al.

Gandal, N. (1994): "Hedonic Price Indexes for Spreadsheets and an Empirical Test of the Network Externalities Hypothesis." *Rand Journal of Economics*, 25, pp. 160-170.

Gandal, N., Kende, M. and Rob (2000): "The Dynamics of Technological Adoption in Hardware/Software Systems: The Case of Compact Disc Players." *Rand Journal of Economics*, 31, pp. 43-61

Gandal, N., Kende, M. and Rob, R. (1997): "The Dynamics of Technological Adoption in Hardware/Software Systems: The Case of Compact Disc Players." Working Paper.

Goerke, L. and Holler, M. J. (1995): "Voting on Standardization." *Public Choice*, 83, pp. 337-351.

Goerke, L. and Holler, M. J. (1998): "Strategic Standardization in Europe: A Public Choice Perspective", *European Journal of Law and Economics*, 6, pp. 95-112.

Goolsbee, A. and Klenow, P.J. (1999): "Evidence on Network and Learning Externalities in the Diffusion of Home Computers." NBER Working Paper #7329.

Gowrisankaran, G. and Stavins, J. (1999): "Network Externalities and Technology Adoption: Lessons from Electronic Payments." Mimeo (University of Minnesota).

Harsanyi, J.C. and Selten, R. (1988): A General Theory of Equilibrium Selection in Games. Cambridge.

Holler M.J. and Simmering, V. (2001): "Voting on Harmonization." Paper presented at the Annual Meeting of the European Public Choice Society, Paris, 2001.

Holler, M.J. (1996): "Modellierung von Netzwerkeffekten und Ansätze Industriepoli-tischer Aussagen." In: Schenk, K.-E., Schmidtchen, D. and Streit, M.E. (eds.): *Jahrbuch für Neue Politische Ökonomie*, 16, pp. 90-114.

Holler, M.J. and Illing G. (1996): Einführung in die Spieltheorie. Berlin, Heidelberg, New York et al.

Holler, M.J. and Peters, R. (1999): "Scandals and Evolution: A Theory of Social Revolution." In: Holler, M.J. (ed.): Scandals and Its Theory. Special issue of *Homo Oeconomicus*, 16, pp. 75-91.

Holler, M.J. and Wickström, B.A. (1999): "The Use of Scandals in the Progress of Society." In: Holler, M.J. (ed.): Scandals and Its Theory. Special issue of *Homo Oeconomicus*, 16, pp. 97-110.

Kahan, M. and Klausner, M. (1995): "Corporations, Corporate Law, and Networks of Contracts." *Virginia Law Review*, 81, pp. 757f.

Kahan, M. and Klausner, M. (1996): "Lockups and the Market for Corporate Control." *Stanford Law Review*, 48, pp. 1539f.

Kahan, M. and Klausner, M. (1997), "Standardization and Innovation in Corporate Contracting (or: 'The Economics of Boilerplate')", *Virginia Law Review*, 83, pp 713f.

Kandori, M. and Rob, R. (1998): "Bandwagon Effects of Long Run Technology Choice." *Games and Economic Behavior*, 22, pp. 30-60.

Kandori, M; Mailath, G.J. and Rob, R. (1993): "Learning, Mutation, and Long Run Equilibria in Games" *Econometrica*, 61, pp. 29-56.

Katz, M.L. and Shapiro C. (1985): "Network Externalities, Competition, and Compatibility." *The American Economic Review*, 75, pp. 424-440.

Katz, M.L. and Shapiro, C. (1986): "Technology Adoption in the Presence of Network Externalities." *Journal of Political Economy*, 94, pp. 822–841.

Katz, M.L. and Shapiro, C. (1992): "Product Introduction with Network Externalities." *The Journal of Industrial Economics*, 40, pp. 55-83.

Katz, M.L. and Shapiro, C. (1994): "Systems Competition and Network Effects." *Journal of Economic Perspectives*, 8, pp. 93-115.

Klemperer, P. (1995): "Competition when Consumers have Switching Costs: An Overview with Applications to Industrial Organization, Macroeconomics, and International Trade." *Review of Economic Studies*, 62, pp. 515-539.

Kohlberg, E. and Mertens, J.-F. (1986): "On Strategic Stability of Equilibria." *Econometrica*, 54, pp. 1003-1037.

Koski, H. (1998): Economic analysis of the Adoption of Technologies with Network Externalities. Diss. Oulo.

Koski, H.A. (1999): "The Installed Base Effect: Some Empirical Evidence From the Microcomputer Market." *Economics of Innovation and New Technology*, 8, pp. 273-310.

Kristiansen, E.G. (1998): "R&D in the Presence of Network Externalities: Timing and Compatibility." *Rand Journal of Economics*, 29, pp. 531-547.

Lange, R., McDade, S. and Oliva, T.A. (2001): "Technological Choice and Network Externalities: A Catastrophe Model Analysis of Firm Software Adoption for Competing Operating Systems." *Structural Change and Economic Dynamics*, 12, pp. 29-57.

Laver, M. and Shepsle, K.A. (1990a): "Coalitions and Cabinet Government." *American Political Science Review*, 84, pp. 873-890.

Laver, M. and Shepsle, K.A. (1990b): "Government Coalitions and Intraparty Politics." *British Journal of Political Science*, 20, pp. 489-507

Lemley, M.A. and McGowan, D. (1998a): "Legal Implications of Network Economic Effects." *California Law Review*, 86, pp. 479f.

Lemley, M.A. and McGowan, D. (1998b): "Could Java Change Everything? The Competitive Propriety of a Propriety Standard." *The Antitrust Bulletin*, pp. 715-773.

Liebowitz, S.J. and Margolis, S.E. (1990): "The Fable of the Keys." *Journal of Law and Economics*, 33, pp. 1-25.

Liebowitz, S.J. and Margolis, S.E. (1994): "Network Externality: An Uncommon Tragedy." *Journal of Economic Perspectives*, 8, pp. 133-150.

MacKie-Mason, J.K. and Varian, H.R. (1995): "Pricing the Internet." In: Kahin, B. and Keller, J. (eds.): Public Access to the Internet. Cambridge.

MacKie-Mason, J.K. and Varian, H.R. (1996): "Some Economics of the Internet." In: Sichel, W. (eds.): Networks, Infrastructure, and the New Task for Regulation. Ann Arbor.

Mailath, G.J. (1992): "Introduction: Symposium on Evolutionary Game Theory." *Journal of Economic Theory*, 57, pp. 259-277.

Mailath, G.J. (1998), "Do People Play Nash Equilibrium? Lessons from Evolutionary Game Theory", *Journal of Economic Literature*, pp. 1347-1374.

Mailath, G.J. (1999): "Evolutionary Game Theory." In: Newman, P. (eds.): *The New Palgrave Dictionary of Economics and the Law*, 2, London, pp. 84-88.

Marburger, P. (1987): "Das Technische Risiko als Rechtsproblem." *Schweizer Ingenieur und Architekt*, 39, pp. 829-838.

Matutes, C. and Regibeau, P. (1988): "Mix and Match: Product Compatibility without Network Externalities." *Rand Journal of Economics*, 19, pp. 221-234.

Matutes, C. and Regibeau, P. (1992): "Compatibility and Bundling of Complementary Goods in a Duopoly." *Journal of Industrial Economics*, 40, pp. 37-54.

Matutes, C. and Regibeau, P. (1996): "A Selective Review of the Economics of Standardization. Entry Deterrence, Technological Progress and International Competition." In Holler, M.J. and Thisse, J.-F. (eds.): The Economics of Standardization. Special issue of *European Journal of Political Economy*, pp. 183–209.

Maynard Smith, J. and Price, G. (1973), "The Logic of Animal Conflict", *Nature*, 246, pp. 15-18.

Meyer, L. and Fontaine, G. (2000): "Final evaluation of the 16:9 Action Plan – Executive summary." Manuscript by Insitute de l'audiovisuel et des telecommunications en europe.

Milgrom, P. and Roberts, J. (1986): "Price and Advertising Signals of Product Quality." *Journal of Political Economy*, 94, pp. 796-821.

Moretto, M. (2001): "Irreversible Investment with Uncertainty and Strategic Behaviour." *Economic Modeling*, forthcoming.

Nurmi, H. (1998): Rational Behavior and the Design of Institutions: Concepts, Theories, and Models. Cheltenham, Northampton.

Peters, R. (1997), "The Stability of Networks: An Evolutionary Approach to Standardization." In Holler, M.J., Niskanen, E. (eds.): *EURAS Yearbook of Standardization*, 1, special issue of *Homo Oeconomicus*, pp. 347-56.

Peters, R. (1998): Evolutionäre Stabilität in sozialen Modellen. (Diss.) München.

Plott, C.R. (1976): "Axiomatic Social Choice Theory: An Overview and Interpretation." *American Journal of Political Science*, 20, pp. 511-596.

Posner, R.A. and Rasmusen, E.B. (1999): "Creating and Enforcing Norms, with Special Reference to Sanctions." *International Review of Law and Economics*, 19, pp. 369-382.

Regibeau, P. and Rockett, K.E. (1996): "The Timing of Product Introduction And the Credibility of Compatibility Decisions." *International Journal of Industrial Organization*, 14, pp. 801-823.

Riker, W.H. (1982): Liberalism Against Populism: A Confrontation Between the Theory of Democracy and the Theory Social Choice. San Francisco.

Rohlfs, J. (1974): "A Theory of Interdependent Demand for Communications Service." *Bell Journal of Economics*, 5, pp. 16-37.

Saari, D.G. (1997): "The Generic Existence of a Core for q-Rules." *Economic Theory*, 9, pp. 219-260.

Saloner, G. and Shepard, A. (1995): "Adoption of Technologies with Network Effects: An Empirical Examination of the Adoption of Automated Teller Machines." *Rand Journal of Economics*, 26, pp. 479-501.

Schäfer, H.B. (2001): "Fixing Statutory Limitation Periods in Sales Contracts." Manuscript (Institute for Law and Economics, University of Hamburg).

Schepel, H. and Falke J. (2000): Legal Aspects of Standardisation in the Member States of the EC and EFTA – Volume 1: Comparative Report. Luxembourg.

Shapiro, C. and Varian, H.R. (1999a): "The Art of Standards Wars." *California Management Review*, 41, pp. 8-32.

Shapiro, C. and Varian, H.R. (1999b): Information Rules: A Strategic Guide to the Network Economy. Boston.

Shy, O. (1996): "Technology Revolutions in the Presence of Network Externalities." *International Journal of Industrial Organization*, 14, pp. 785-800.

Shy, O. (2001): The Economics of Networks. Cambridge, New York, Oakleigh et al.

Simmering, V. (2000): "The Diffusion of Network Technologies." In: Coenen, H.; Holler, M.J. and Niskanen, E. (eds.): *Discussion papers of 5th Helsinki Workshop on Standardization and Networks.* VATT, Helsinki.

Störig, H.J. (1992): Abenteuer Sprache. München.

Sykes, A.O. (1995): Product Standards for Internationally integrated goods markets. Washington.

Thum (1994): "Möglichkeiten und Grenzen staatlicher Standardsetzung." In: Tietzel, M. (ed.): Ökonomik der Standardisierung. Special issue of *Homo Oeconomicus*, 11, pp. 465-99

Thum, M. (1994): "Network Externalities, Technological Progress, and the Competition of Market Contracts." *International Journal of Industrial Organization*, 12, pp. 269-289.

Thum, M. (1995): Netzwerkeffekte, Standardisierung und staatlicher Regulierungsbedarf. Diss. Tübingen.

Voelzkow, H. (1996): Private Regierungen in der Techniksteuerung: eine sozialwissenschaftliche Analyse der technischen Normung. Frankfurt/Main.

Weitzman, M. (1983): "Contestable Markets: An Uprising in the Theory of Industry Structure: Comment." *American Economic Review*, 73, pp. 486-487.

Wiese, H. (1997): "Compatibility, Business Strategy and Market Structure – A Selective Survey." In: Holler, M.J. and Niskanen, E. (eds.): *EURAS Yearbook of Standardization*, 1, special issue of *Homo Oeconomicus*, pp. 283-308.

Woeckener, B (1997): "The European Standardization System: How Much in Need of Reform Is It?" In: Holler, M.J. and Niskanen, E. (eds.): *EURAS Yearbook of Standardization*, 1, special issue of *Homo Oeconomicus*, pp. 391-410.

Young, H.P. (1993): "The Evolution of Conventions." *Econometrica*, 61, pp. 57-84.